Asymptomatic Osseous Variations of the Postcranial Human Skeleton

Comparative Anatomy and Posture of Animal and Human Set

coordinated by
Djillali Hadjouis

Volume 5

Asymptomatic Osseous Variations of the Postcranial Human Skeleton

Emeline Verna

iSTE

WILEY

First published 2023 in Great Britain and the United States by ISTE Ltd and John Wiley & Sons, Inc.

ISTE Ltd
27-37 St George's Road
London SW19 4EU
UK

www.iste.co.uk

John Wiley & Sons, Inc.
111 River Street
Hoboken, NJ 07030
USA

www.wiley.com

Library of Congress Control Number: 2022949595

British Library Cataloguing-in-Publication Data
A CIP record for this book is available from the British Library
ISBN 978-1-78630-880-1

Contents

Foreword

Asymptomatic Osseous Variations of Postcranial Human Skeleton is the result of doctoral research whose quality was unanimously praised by the members of the jury and addresses a little-known subject with forensic applications, but also with implications for archaeological studies.

Anatomical variations have been known for a long time, but their study has been neglected by anthropologists

As a result, today, few works present even an inventory of these variations and, when it is the case, it is mostly limited to the cranial skeleton. It was therefore time for a study to take stock of this subject, including an inventory of the other anatomical regions of the skeleton.

In this work, Emeline Verna proposes an anatomical definition of the most relevant asymptomatic bone variations according to the context of study (forensic or archeological), then an estimation of the frequency of these variations, their sexual dimorphism, their lateralization and their variability based on the study of two populations: one contemporary, the other osteo-archaeological, and completed by data from the literature.

This work, which was lacking until now, can be used as a document in the study of anatomical variations and as a reference for postcranial variations.

The quality of the illustrations accompanied by multiple frequencies for each variation will allow the researcher to use it in the course of their research or expertise. This book presents an approach at the interface of several disciplines, resulting from fundamental research, but also from applied cases, in a perspective of individual identification and thus forensic, as well as in a population approach and thus a better knowledge of past populations. Anatomists, anthropologists, forensic scientists and radiologists will undoubtedly find it a valuable aid.

Jaroslav BRŮŽEK
Professor at the Faculty of Science
Charles University, Prague
Director of Research Emeritus, UMR PACEA

Preface

This book enables, in a simple and quick way, using asymptomatic osseous variations (AOVs) in the different research areas of biological anthropology. It will allow everyone to use it according to their study objective. Some AOVs have a different frequency depending on the population and some depending on the study medium (dry bone, imaging). It is also interesting to know which ones have a very high frequency (above 80–90%), the observation of the absence of the variation being of interest.

A good knowledge of all the AOVs found on the post-cranial skeleton will facilitate their use. Even if the exact etiology of the majority of them remains unknown, a clear definition, coupled with good illustrations and access to a population frequency, will enable a simple and quick use in different contexts. Whatever the context, their use can provide information on individuals and thus better characterize them.

This book presents each of the 82 AOVs present on the human postcranial skeleton and brings together different information:

– synonyms in English and Latin;

– a simple description that enables placing the AOV on the bone and knowing its origin;

– an illustration based on four images: an Illustrator diagram, on dry bone (osteo-archaeological collections of the UMR 7268 or the Muséum national d'Histoire naturelle de Paris), a 3D reconstruction using Avizo software and an image of a section in DICOM format (medical imagery of the Assistance publique des hôpitaux de Marseille);

– a frequency based on dry bone studies (osteological collections) and/or a frequency based on medical imaging studies (X-ray, Magnetic Resonance Imaging (MRI), computerized tomography (CT) scan) depending on the population studied;

– observed associations for AOVs by sex, age and laterality (highlighted using statistical tests).

Regarding illustrations, all four media are not always available for each of the AOVs. Some variations, especially those of the upper limb, are more rarely observed from imaging (apart from radiography, the arm is rarely scanned, meaning this book has limited access to images). All of the illustrations were collected by the author, which sometimes explains the lack of certain images that were not found during the study.

December 2022

Acknowledgments

This book would not have been possible without the contribution of several people whom I wish to thank.

First of all, I would like to thank my thesis supervisors, Professor Pascal Adalian and Professor Marie-Dominique Piercecchi-Marti, who have guided me toward the next step of my career.

I thank Professor Kathia Chaumoitre and Professor Michel Panuel (Assistance publique des hôpitaux de Marseille) for allowing me access to the medical imaging collection in connection with my doctoral research work; the UMR 7268 and its director for the access to the osteological collections; and the persons in charge of the osteological collections of the Muséum national d'Histoire naturelle de Paris.

I would like to thank the French Ministry of Higher Education, Research and Innovation for the funding of the research grant that allowed me to carry out this work during my university doctorate.

I would also like to thank Professor Jaroslav Brůžek for writing the foreword to this book.

And, of course, I thank Dr. Caroline Costedoat, Dr. Sandy Parmentier and Dr. Michel Signoli for their unfailing support.

Introduction

Anatomical variations have long been of interest to anthropologists and anatomists, in particular, since the end of the 19th century. Even if the first description of anatomical variations found in the literature dates back to antiquity with the description of wormian bones made by Hippocrates (Le Double 1903), it was not until the end of the 19th century and the beginning of the 20th century that specific studies on these variations appeared. These works were then only purely descriptive and were mainly carried out by anatomists, which made it possible to compile a substantial body of data. The works of Le Double (1903, 1906, 1912), himself an anatomist, are good examples, and his treatises are still authoritative in the medical field. In particular, he wrote important works on the vertebrae, grouping together all the so-called "abnormal" variations of the axial skeleton (Le Double 1912). Other studies turned to a comparative approach, such as that of Chambellan (1883). This was the first use of anatomical variations as anthropological characters for the purpose of comparison between individuals or populations (Gemmerich 1999). However, this work is primarily used to validate hierarchical social and racial classifications, and the comparison of individuals to each other has led to the definition of "racial diagnostic criteria" (Wood-Jones 1931, 1934). The study of anatomic variation was generally neglected between the two world wars, but then gained momentum in the 1950s under the impetus of researchers such as Berry and Berry (1967) and Ossenberg (1969). At that time, studies shifted from an intra-population to an inter-population approach in order to study the affinity between human groups (Berry 1974; Murail 2005), that is, microevolutions or settlement dynamics. First, they are carried out from a sample of cranial variations to make a simple anatomical comparison of populations, then, in a second step, they highlight processes of settlement and affinity between populations (combination of several types of data: anthropological, genetic, linguistic, geographical, etc.). These advances are to be placed in parallel with the command of new statistical tools allowing, among other things, the estimation of biological distances (Grüneber 1952) by the "mean measure of divergence" (MMD) (revised by Sjovold 1973).

Anatomical variations have also been used in phylogenetic analyses of hominid fossils (Thoma 1981) or of present-day great apes (Braga 1993). The results are identical to those obtained from genetics and thus show the potential of the latter. Experiments carried out in known situations have made it possible to highlight relationships between different human groups, to show certain evolutions within groups, but also to study the genetic determinism and to calculate the rate of heritability of certain variations (Berry 1975; Saunders and Popovitch 1978; Cheverud and Buikstra 1981a, 1981b, 1982; Sjovold 1984). The influence of environmental factors has also been studied (Bocquet-Appel 1984). All these studies have made it possible to qualify the influence given to anatomical variations as "family markers" and to show that the appearance of these variations is subject to both genetic and environmental factors (Fily et al. 2001). The conclusions obtained from their use in the context of funerary anthropology must be qualified and formulated more in the form of hypotheses. Synthesis work is also being developed to bring together as much variation as possible to serve as a baseline for the skull, teeth and postcranial skeleton (Hauser and De Stefano 1989; Saunders 1989; Scott and Turner 1997).

In recent years, the study of these variations has focused mainly on funerary anthropology. Crubézy (1990, 1991, 1992) has used discrete traits in studies mixing anatomy and genetics in order to highlight possible family groupings, while keeping in mind that only biological kinship, and not social kinship, can be understood. It therefore seems more appropriate to speak of how biological proximity within a funerary group Gemmerich (1999) tested the relationship between kinship and anatomical variation on a Swiss reference collection (Simon collection). No correlation was found, thus calling into question the use of anatomical variations as "family markers". However, this study did highlight sociodemographic behaviors and moved toward the interpretation of microevolutionary phenomena (Murail 2005). This has been confirmed by the study of the Missiminia necropolis (Sudan) (Crubézy et al. 1999), which made it possible to study population movements and the organization of burial spaces over a period of three centuries based on anatomical variations.

Studies on DNA analysis, especially on mitochondrial DNA (Passarrius et al. 2008), have been undertaken to confirm the possibility of the relationship between individuals and thus validate the genetic transmission of certain variations (Gruneberg 1952). However, the DNA must be well preserved and uncontaminated, which limits the conclusions.

These variations are also used in settlement problems in Switzerland for the Campaniform period (Desideri and Eades 2002), America (Ossenberg 1994) and Polynesia (Hanihara 1992). They are also used in the classification of human groups (Hanihara et al. 2003).

Finally, they can be used in forensic anthropology to assist in the identification of individuals (Verna et al. 2014, 2015; Verna 2016). The principle is to improve the characterization of the individual's biological profile by means of scoring rare anatomical variations in order to advance toward an ante-/post-mortem comparative identification, or even a positive identification (Riddick 1983; Schmitt et al. 2006). Indeed, in forensic anthropology, but also more broadly in biological anthropology, the determination of sex, the estimation of age at death and the estimation of stature are essential parameters in the constitution of an individual's biological profile from skeletonized remains. However, the construction of the latter from these three parameters alone limits the interpretations, because several individuals may share the same profile. Moreover, the methods of estimating age and stature are still imprecise, especially for adult individuals, and give, in the majority of cases, wide ranges of estimation that can correspond to several individuals. As for the determination of sex from skeletonized remains, it is not always possible to determine it with certainty if the coxal bone (the most discriminating bone) is missing. Without these data or with imprecise data, the contribution of anthropological expertise remains limited. The introduction and observation of other parameters such as pathologies or any other distinctive sign, notably anatomical variations, can be an asset in establishing the most complete and singular biological profile possible in order to correspond to a very limited number of individuals or, in the best case, to a single individual (Schmitt et al. 2006). In addition to anatomical variations, the observation of postural markers and congenital anomalies can also be an interesting way to complete the biological profile. However, before any use, a definition of each entity is necessary in order to better identify their characteristics and their interests.

Anatomical variations are better known as discrete traits. These are numerous, inconsistent and asymptomatic anatomical traits found on the majority of bones (Saunders 1978). The appearance of discrete traits follows the all-or-nothing law (marked as absent or present), but for some of them different stages may be appreciable (especially cranial discrete traits). Scoring by stages, however, leads to many intra- and inter-observer errors (Murail 2005). All-or-nothing scoring, 0 for absence of the character and 1 for presence, is preferred for the postcranial skeleton. Some characters can be more or less voluminous (such as the third trochanter) (Dastugue 1984) and be measurable. The discrete/measurable opposition does not necessarily exclude the character as "discrete"; it depends on the orientation that one wishes to give to the study. One must only be careful when using them. Concerning the frequency of these traits in different populations, it is specific to each one and can vary according to the sex, the side and the age. An inter-population difference, sometimes important, can be observed for some traits. In the literature, a tendency is emerging to study more discrete traits of the skull (Hauser and De Stefano 1989) at the expense of those present on the postcranial skeleton. The skull has long been considered the centerpiece of the skeleton (Gemmerich 1999). Discrete traits on the postcranial skeleton have received less attention, although there has been some synthetic work on this region, such as that of

Saunders (1978) and Finnegan (1978). The origin of discrete traits is complex and has been debated since the beginning of their use. A single determinism common to all is not possible (Murail 2005). Their heritability seems to be rather low, even if there are some exceptions identified through the study of family cases (Saunders and Popovich 1978), but these do not respond to the simple laws of genetics, that is, one gene for one trait (Murail 2005). Poly-factorial determinism is therefore to be preferred and animal experiments (Grüneberg 1952) support this hypothesis. Some epigenetic factors may also intervene in the appearance of these traits (such as deficiencies during childhood). The origin of discrete traits is polyfactorial, depending on a particular gene pool and environment. It is therefore more prudent to make interpretations at the population level and not at the individual level in the case of studies of funerary groups or settlements.

Congenital anomalies can be defined as structural bone variations deviating from standard human morphology, that is, abnormalities of structure or function. These anomalies may be pathological or non-pathological (Barnes 2012) and have a genetic origin (controlled by a gene pool). The major work on congenital anomalies is Ethne Barnes' *Atlas of Developmental Field Anomalies of the Human Skeleton: A Paleopathology Perspective* (2nd edition 2012). The latter references all congenital anomalies of the human skeleton. In this handbook, only anomalies qualified as asymptomatic have been retained. They may be more or less frequent depending on the population, with some having no frequency at all in certain populations (Barnes 2012). This is the case, for example, for the butterfly vertebra (Merbs 2004). These differences provide insight into the genetic pattern of the population, that is, family relationships, intermarriage and migration (Barnes 2012). These anomalies allow for the characterization of an individual, knowing that they develop during embryogenesis and are therefore present from an early age.

Skeletal markers of activity are symptomatic or asymptomatic variations representing the response of the bone to the different stresses experienced during the individual's life (Kennedy 1989). These markers can be classified into three groups (Villotte 2008) according to the nature of the bone manifestation, taking into account the extreme variability of these markers: structural skeletal adaptations, pathological changes and supernumerary facet joints and other discrete traits. In the following, only markers belonging to the group of supernumerary facet joints will be considered because of their asymptomatic nature, and we will group them under the term postural markers. Indeed, they cannot be qualified as discrete traits, as they are the consequence of biomechanical factors. They were first described in the medical literature in the middle of the 17th century as diseases related to work (especially mining). Then, from the 18th to 19th century, they were considered as the effects of work on the human skeleton (described by Charles Turner Thackrah (1795–1833) and William Arbuthnot Lane (1856–1943); see Kennedy 1989). At the end of the 19th century, an important synthesis was made by William Turner bringing together a large number of the activity markers (Kennedy 1989). Aleš Hrdlička, at the beginning of the 20th century, paid particular attention to these markers, and many studies followed his discoveries. An important work in the field is Fransceco Ronchese's *Occupational Marks and Other Physical Signs: A Guide to Personal Identification* published in 1948 (Capasso et al. 1999). Currently, the most comprehensive synthesis of 130 activity-related skeletal markers is the *Atlas of Occupational Markers on Human Remains* by Capasso et al.

Posture markers are used in many fields of biological anthropology (paleoanthropology, bioarcheology, forensic anthropology, etc.) and they propose interpretations on the activities of past populations in order to better understand their lifestyle (Villotte 2009). Reconstructions of activity models based on current experiments with known activities (Dutour 1992, 2000) have been carried out in an attempt to understand the association between a marker and an activity. In our approach, only asymptomatic posture markers will be selected. Enthesopathies will not be used because of their pathological nature, even if previous studies sometimes associate these enthesopathies with the study of discrete traits (Finnegan 1978; Saunders 1978).

Dastugue (1984) proposes a definition of a discrete trait as "a sign" and its essential, primary quality is to be maximally "meaningful". This definition seems to be equally applicable to the congenital anomalies and posture markers selected in our study.

To be "meaningful", they must meet several conditions:

– They must be unambiguously definable: this is relatively complicated for some of the discrete traits, as it is difficult to find a consensus definition in the literature, most of them having multiple definitions more or less close to each other (e.g. femoral neck region).

– They must be constitutional and normal: these are two difficult states to fulfill. The limit between constitutional and acquired is sometimes very delicate to grasp for certain characteristics (e.g. the squat facets of the tibia and the talus), which cannot then be considered as discrete traits in the strict sense of the term. However, these are neither measurable nor symptomatic, illustrating the difficulty of defining a discrete trait in the strict sense. This remark also applies to certain congenital anomalies (*spina bifida occulta*). Similarly, for the term "normal", what is normality? It seems difficult to mark a difference between the normal and the abnormal (Dastugue 1982) and it is for this reason that the study of discrete traits very often rubs shoulders with the pathological with a sometimes very thin limit between these two worlds.

– They must establish a possible hierarchy: this should depend on the purpose and material of the study. It is not necessary to take all discrete cranial, dental and postcranial features for the study. For example, in forensic anthropology, only those with a frequency of less than 10% and that are easily observable in an IR image can be considered relevant and therefore selected.

– They must be stable: that is to say, studies on mature adult individuals should be favored in order to avoid the constraints linked to bone maturation which could prevent the preservation of certain characteristics.

This approach allows us to give a common definition and to orient the choice of variations according to the objective of the study.

To facilitate the use of these different terms (discrete traits, asymptomatic congenital anomalies and postural markers), they are grouped under a single name: *asymptomatic osseous variations* (AOVs).

The definition of variation may differ somewhat between authors, leading to confusion and misuse of a name. In this book, all the synonyms found in the literature for each of the AOVs are presented in order to avoid confusion and to be able to rely on a clear definition.

As few classifications have been made for the postcranial skeleton; we have taken the classification of discrete cranial features established by Ossenberg (1969), while adapting it and including in the best possible way congenital anomalies and posture markers. The proposed classification is thus as follows:

– hyperostotic variations: variations resulting from the ossification of cartilaginous or ligamentous elements and, by extension, any variation producing an over-ossification compared to the original state;

– hypostotic variations: variations resulting from incomplete ossification or persistence of immature stages in adulthood;

– supernumerary bones;

– facet joint variations: duplication, absence, position and number.

1

The Upper Limb

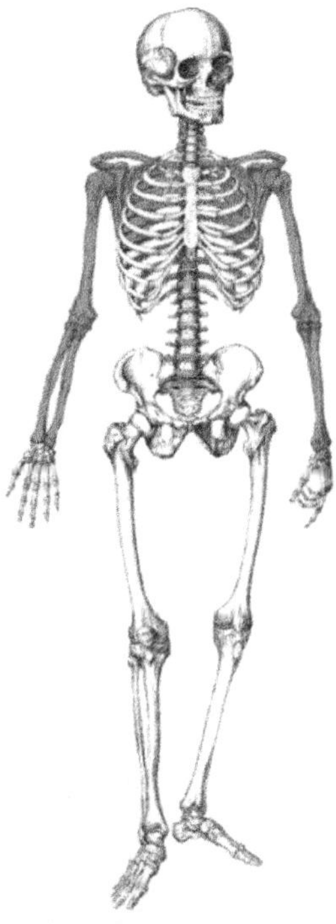

Figure 1.1. *Human skeleton. The upper limb is presented in red.*
For a color version of this figure, see www.iste.co.uk/verna/asymptomatic.zip

This chapter focuses on the asymptomatic osseous variations (AOVs) present on the upper limb. The enumeration of these AOVs is done by bone in order to be able to refer to them easily and according to the objective of the study. This part of the body is very rarely scanned and more often X-rayed in the case of living beings, which explains the absence of medical imaging for many AOVs.

The hand has a large number of bones, but with little associated variation; therefore, this element has not been addressed in this book.

1.1. Clavicle

The clavicle is composed of two primary and two secondary ossification centers (Scheuer and Black 2004; Scheuer et al. 2009). Fusion of the lateral epiphysis occurs around 18–20 years of age; the medial epiphysis fuses around 20–30 years of age. This is the epiphysis of the human skeleton that fuses the latest.

All the AOVs described for the clavicle can be observed from a very young age. Skeletal maturation does not influence their observation. None of these AOVs belong to the hypostotic category.

A total of seven AOVs are presented in this book for the clavicle.

1.1.1. *Perforation of the clavicle*

1.1.1.1. *Definition*

– Synonyms: Supraclavicular nerve foramen and supraclavicular nerves pierce clavicle.

– Description: It is located at the boundary between the anterior edge and the upper surface of the clavicle and may allow the supraclavicular nerve to pass through.

1.1.1.2. *Illustrations*

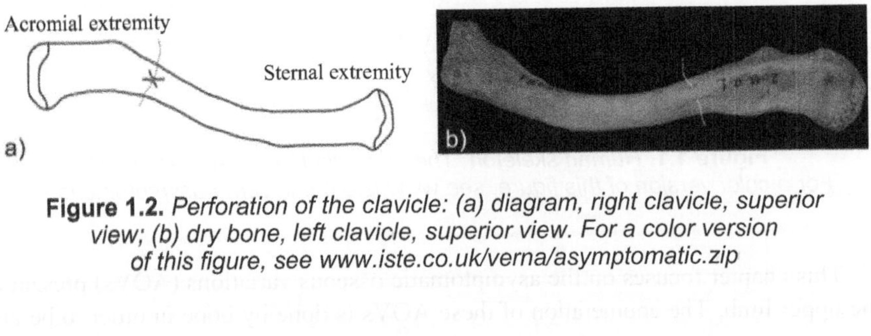

Figure 1.2. *Perforation of the clavicle: (a) diagram, right clavicle, superior view; (b) dry bone, left clavicle, superior view. For a color version of this figure, see www.iste.co.uk/verna/asymptomatic.zip*

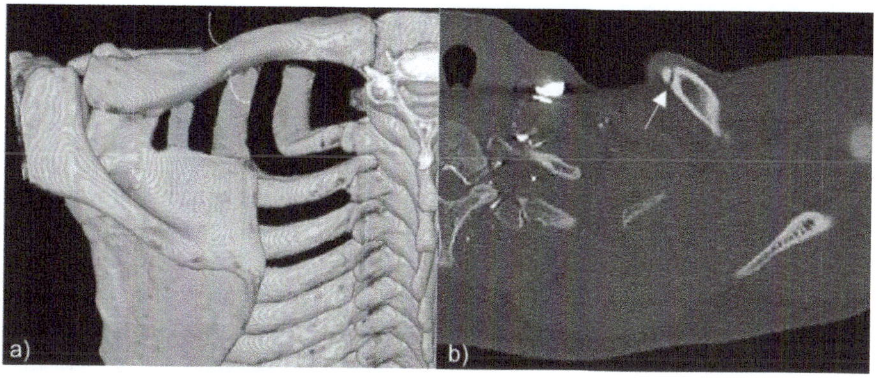

Figure 1.3. *(a) 3D reconstruction, left clavicle, postero-superior view; (b) DICOM image, left clavicle, axial section. For a color version of this figure, see www.iste.co.uk/verna/asymptomatic.zip*

1.1.1.3. *Population frequencies*

Frequency	Population	Reference
1.3	Mediterranean	(Verna 2014)

Table 1.1. *Frequency of perforation of the clavicle in medical imaging*

Frequency	Population	References
0	American (17th–18th centuries)	(Saunders 1978)
0.4	American (Eskimo, 16th–18th centuries)	(Saunders 1978)
1.9	American (8th–10th centuries)	(Saunders 1978)
2.3	French (4th–8th centuries)	(Debono 2006)
2.2	Greek (21st century)	(Natsis et al. 2016)
7.4	French (9th–12th centuries)	(Verna 2014)
8.1	French (19th–20th centuries)	(Verna 2014)

Table 1.2. *Frequency of perforation of the clavicle on dry bone*

1.1.1.4. *Association(s)*

This AOV is more often observed unilaterally than bilaterally.

1.1.2. *Coracoclavicular joint*

1.1.2.1. *Definition*

– Synonym: Subclavian facet.

– Description: It is located between the clavicle and the coracoid process of the scapula. It results in a facet joint at the insertion of the conoid ligament.

1.1.2.2. *Illustrations*

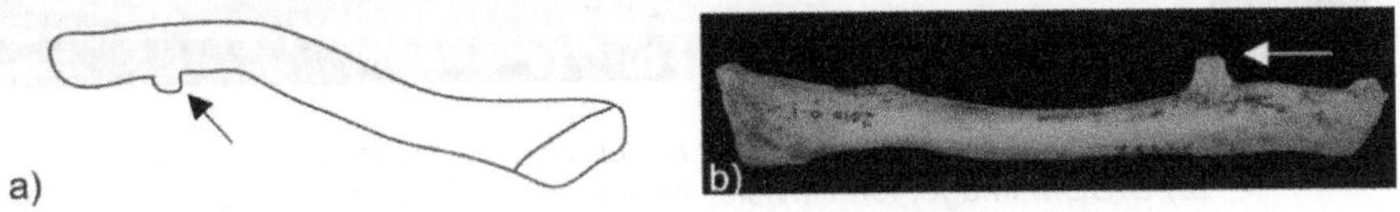

Figure 1.4. *Coracoclavicular joint: (a) diagram, right clavicle, lateral view; (b) dry bone left clavicle, lateral view. For a color version of this figure, see www.iste.co.uk/verna/asymptomatic.zip*

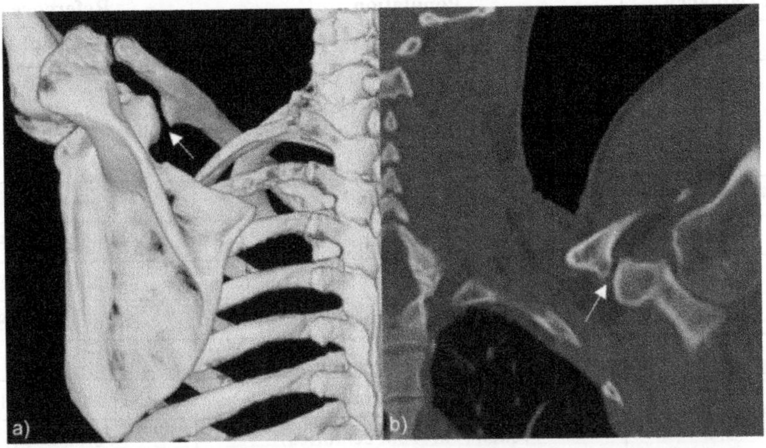

Figure 1.5. *(a) 3D reconstruction, left clavicle, posterior view; (b) DICOM image, left clavicle, coronal section. For a color version of this figure, see www.iste.co.uk/verna/asymptomatic.zip*

1.1.2.3. *Population frequencies*

The highest frequencies were found in Eskimo (Saunders 1978) or Neolithic populations (Thomas 2011).

An observation put forward by Voisin (2012) highlights a difference between the frequencies obtained from imaging and those from dry bone; it would seem lower for the latter. This trend is not found here since the highest frequency is observed on dry bone.

The latter may be explained by an overrepresentation of individuals over 50 years of age in the French collection dating from the 20th century (Verna 2014), then the formation of this joint occurred following trauma that may have increased the frequency somewhat.

Frequency	Population	References
1.2	Dutch	(Haas et al. 1965)
1.6	Mediterranean	(Verna 2014)

Table 1.3. *Frequency of the coracoclavicular joint in medical imaging*

Frequency	Population(s)	Reference(s)
0.2	American (17th–18th centuries)	(Saunders 1978)
1.9	French (9th–12th centuries)	(Verna 2014)
5.6	Indian (20th century)	(Das et al. 2016)
6.2	American (8th–10th centuries)	(Saunders 1978)
6.6	French (19th–20th centuries)	(Verna 2014)
12.3	American (Eskimo, 16th–18th centuries)	(Saunders 1978)
14	French (Neolithic)	(Thomas 2011)

Table 1.4. *Frequency of the coracoclavicular joint on dry bone*

1.1.2.4. *Association(s)*

No association was found for this AOV.

1.1.3. *Duplication of the acromial extremity*

1.1.3.1. *Definition*

– Synonym: Duplication of the clavicle.

– Description: The acromial extremity of the clavicle may have a double extremity with two separate joints.

Two origins have been suggested: traumatic (Wilson 2006) and congenital (Golthamer 1957).

1.1.3.2. *Illustrations*

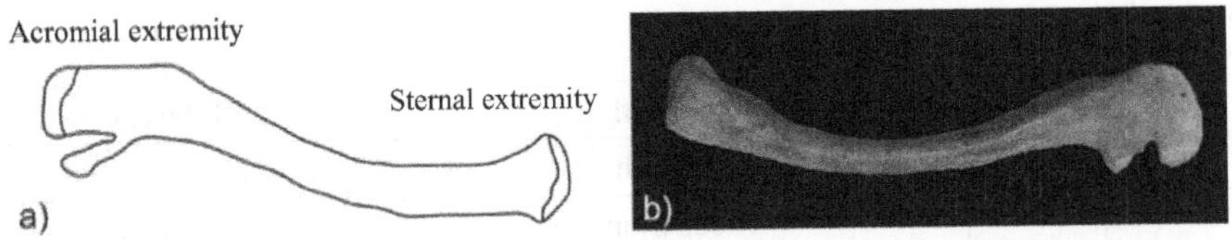

Figure 1.6. *Duplication of the acromial extremity of the clavicle: (a) diagram, right clavicle, superior view; (b) dry bone, left clavicle, superior view. For a color version of this figure, see www.iste.co.uk/verna/asymptomatic.zip*

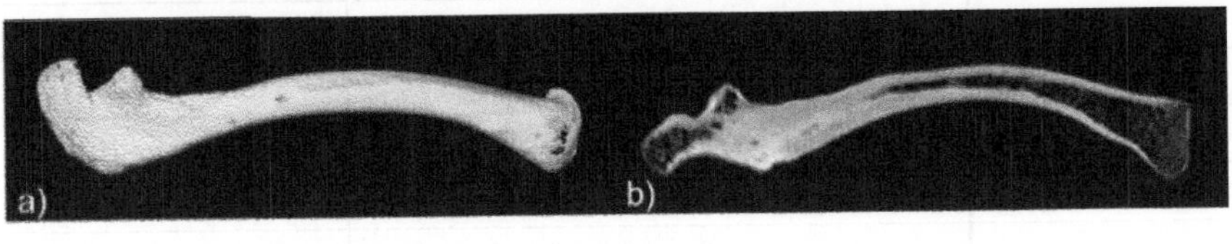

Figure 1.7. *(a) 3D reconstruction, left clavicle, superior view; (b) DICOM image, left clavicle, superior view*

1.1.3.3. *Population frequencies*

Only case reports (Golthamer 1957; Twigg 1981; Wilson 2006; Verna et al. 2017) relate its presence in the literature reflecting the rarity of extremity duplication. The latter may be the result of an old fracture (Wilson 2006) and, in this case, it is not considered as an AOV, but as a pathology.

Frequency	Population	References
0	Mediterranean	(Verna 2014)
Case report	American	(Twigg and Rosenbaum 1981)

Table 1.5. *Duplication frequency of the acromial extremity in medical imaging*

Frequency	Population	References
0	French (19th–20th centuries)	(Verna 2014)
0	French (9th–12th centuries)	(Verna 2014)
Case report	French (19th century)	(Verna et al. 2017)

Table 1.6. *Duplication frequency of the acromial extremity on dry bone*

1.1.3.4. *Association(s)*

No association was found for this AOV.

1.1.4. *Costoclavicular sulcus*

1.1.4.1. *Definition*

– Synonyms: Its synonyms are insertion of the rhomboid pit, rhomboid fossa and costal impression.

Description: It is located at the point of attachment of the costoclavicular ligament, at the sternal extremity, on the inferior surface of the clavicle as a rough, oval depression or tubercle.

1.1.4.2. *Illustrations*

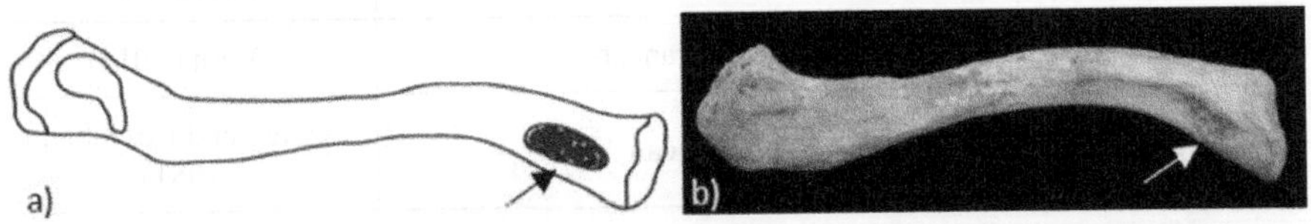

Figure 1.8. *Costoclavicular sulcus: (a) diagram, left clavicle, inferior view; (b) dry bone, left clavicle, inferior view. For a color version of this figure, see www.iste.co.uk/verna/asymptomatic.zip*

1.1.4.3. *Population frequencies*

Frequency	Population	References
No study		

Table 1.7. *Frequency of the costoclavicular sulcus in medical imaging*

Frequency	Population	References
3.2	American (8th–10th centuries)	(Saunders 1978)
9.3	American (17th–18th centuries)	(Saunders 1978)
17.2	American (Eskimo, 16th–18th centuries)	(Saunders 1978)
23	French (4th–5th centuries)	(Castex 1994)
25.3	French (19th–20th centuries)	(Verna 2014)
40	French (Neolithic)	(Thomas 2011)
45	French (4th–6th centuries)	(Castex 1994)
45	French (5th–6th centuries)	(Castex 1994)
47.5	French (9th–12th centuries)	(Verna 2014)
50	French (12th century)	(Castex 1994)
53.3	French (8th–12th centuries)	(Crubézy 1988)
55	French (5th–7th centuries)	(Castex 1994)
57.8	Indian (20th century)	(Jit and Kaur 1986)

Table 1.8. *Frequency of the costoclavicular sulcus on dry bone*

This AOV presents a very variable frequency and is therefore population dependent. Significant differences are visible within the same population (French or American population, for example), which can be related to the historical period.

The more contemporary the population, the lower the frequency. The hypothesis of a less important solicitation of the shoulder during a child/adolescent's development can be proposed.

1.1.4.4. *Association(s)*

This variation is predominantly present in men and on the right side in all populations.

It is also more often found bilaterally than unilaterally.

1.2. Scapula

The scapula ossifies from a primary ossification center. It has three secondary ossification centers for the coracoid process, one for the acromion, two for the glenoid cavity and four for the medial border.

The medial border fuses to the body of the scapula around 19–23 years of age, as does the inferior angle. The acromion fuses to the rest of the scapula around 18–20 years (Scheuer and Black 2004; Scheuer et al. 2009). On the scapula, six AOVs can be observed.

Please note that one variation – the acromial bone – is considered hypostotic, that is, it can only be recorded as a variation from the age of 20 (skeletal maturation completed), and the other five can be observed from a very young age.

1.2.1. *Coracoclavicular joint*

1.2.1.1. *Definition*

– Synonyms: None.

– Description: It is located between the clavicle and the coracoid process of the scapula.

It results in a facet joint at the insertion of the conoid ligament.

1.2.1.2. *Illustrations*

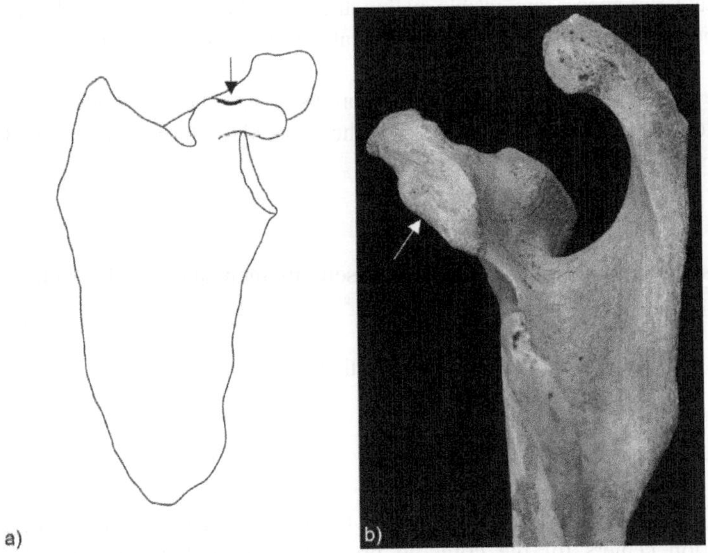

Figure 1.9. *Coracoclavicular joint: (a) diagram, left scapula, anterior view; (b) dry bone, right scapula, lateral view. For a color version of this figure, see www.iste.co.uk/verna/asymptomatic.zip*

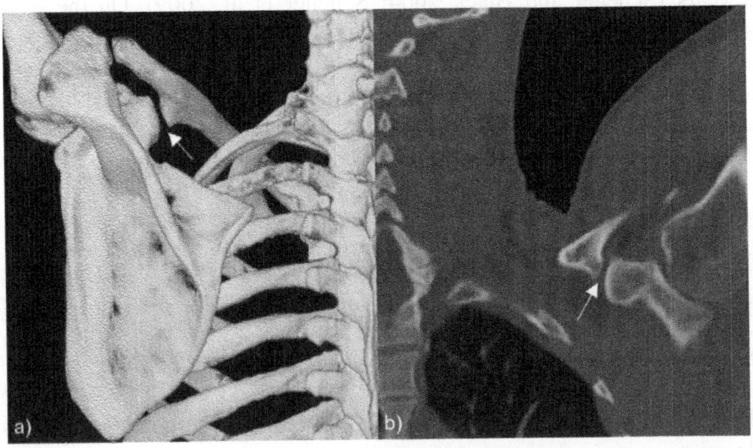

Figure 1.10. *(a) 3D reconstruction, left scapula, posterior view; (b) DICOM image, left scapula, coronal section. For a color version of this figure, see www.iste.co.uk/verna/asymptomatic.zip*

1.2.1.3. *Population frequencies*

Few literatures propose a frequency for this variation; it is implicitly associated with that of the clavicle. It varies from 0.2 to 2%.

This joint is more like a depression in response to the clavicle joint, creating a contact between the two parts.

Frequency	Population	References
0.2	Mediterranean	(Verna 2014)
1.2	Dutch	(Haas et al. 1965)

Table 1.9. *Frequency of the coracoclavicular joint in medical imaging*

Frequency	Population	References
2.1	French (19th–20th centuries)	(Verna 2014)

Table 1.10. *Frequency of the coracoclavicular joint on dry bone*

1.2.1.4. *Association(s)*

No association was found for this AOV.

1.2.2. *Suprascapular foramen*

1.2.2.1. *Definition*

– Synonyms: None.

– Description: It is located at the junction of the coracoid process on the superior border due to the complete ossification of the suprascapular ligament.

1.2.2.2. *Illustrations*

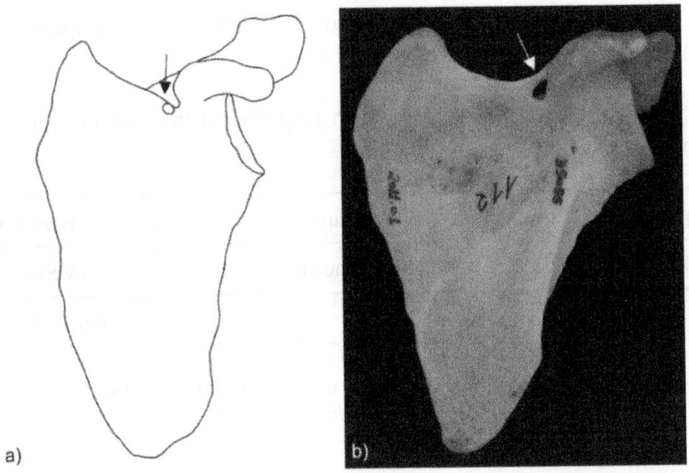

Figure 1.11. *Suprascapular foramen: (a) diagram, left scapula, anterior view; (b) dry bone, left scapula, anterior view. For a color version of this figure, see www.iste.co.uk/verna/asymptomatic.zip*

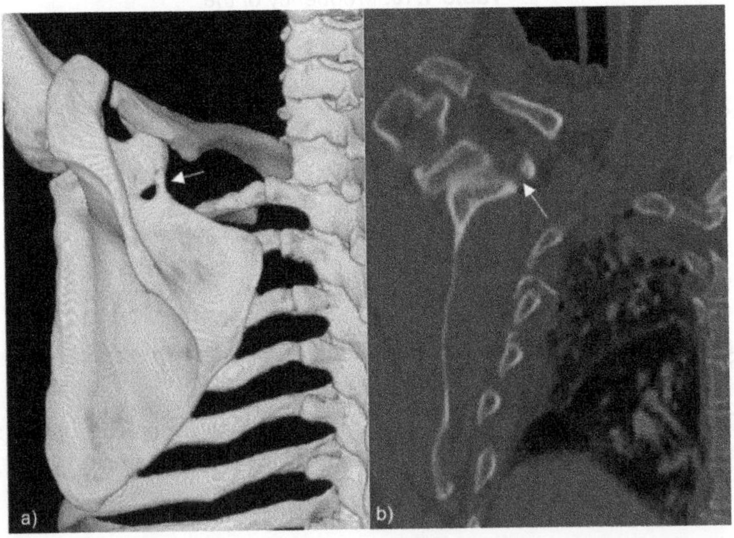

Figure 1.12. *(a) 3D reconstruction, left scapula, posterior view; (b) DICOM image, right scapula, coronal section. For a color version of this figure, see www.iste.co.uk/verna/asymptomatic.zip*

1.2.2.3. *Population frequencies*

Frequency	Population	References
4.2	Mediterranean	(Verna 2014)
5.2	Polish	(Polguj et al. 2012)

Table 1.11. *Frequency of suprascapular foramen in medical imaging*

Frequency	Population	References
0	American (8th–10th centuries)	(Saunders 1978)
0.8	American (Eskimo, 16th–18th centuries)	(Saunders 1978)
3.1	Indian (21st century)	(Saikia et al. 2017)
3.8	American (17th–18th centuries)	(Saunders 1978)
5.1	French (19th–20th centuries)	(Verna 2014)
6.3	American (20th century)	(Gray 1942)
7	Polish (anatomical collection)	(Polguj et al. 2011)
7.3	German (20th century)	(Natsis et al. 2007)

Table 1.12. *Frequency of suprascapular foramen on dry bone*

1.2.2.4. *Association(s)*

This AOV is observed more often unilaterally than bilaterally. Moreover, its frequency tends to increase with age (over 60 years).

This can be explained by late ossification of the suprascapular ligament.

1.2.3. *Suprascapular notch*

1.2.3.1. *Definition*

– Synonyms: None.

– Description: It is located on the upper edge at the junction with the coracoid process.

1.2.3.2. *Illustrations*

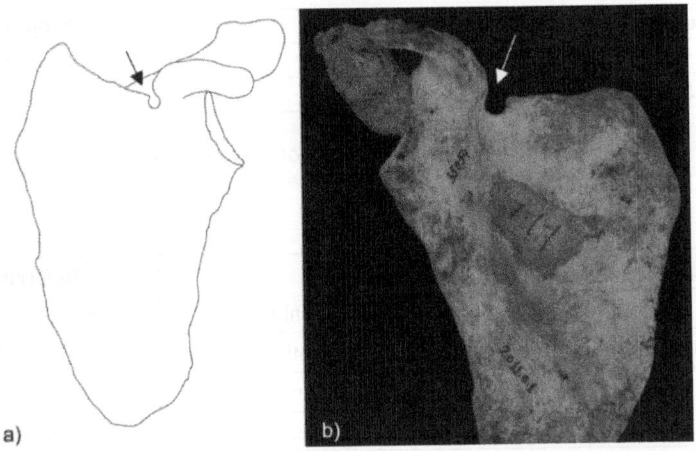

Figure 1.13. *Suprascapular notch: (a) diagram, left scapula, anterior view; (b) dry bone, right scapula, anterior view. For a color version of this figure, see www.iste.co.uk/verna/asymptomatic.zip*

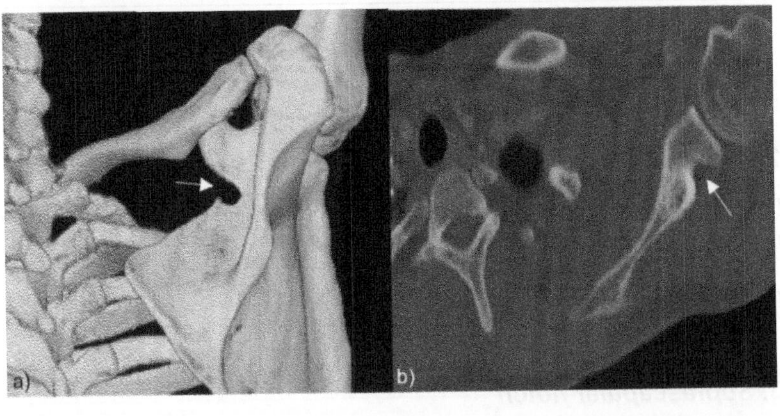

Figure 1.14. *(a) 3D reconstruction, right scapula, posterior view; (b) DICOM image, left scapula, axial section. For a color version of this figure, see www.iste.co.uk/verna/asymptomatic.zip*

1.2.3.3. *Population frequencies*

The frequency of suprascapular notch is high, ranging from 58% to 87% in all populations. The absence of the suprascapular notch seems more interesting to us than its presence.

Frequency	Population	References
71.9	Mediterranean	(Verna 2014)
87.2	Polish	(Polguj et al. 2012)

Table 1.13. *Frequency of the suprascapular notch in medical imaging*

Frequency	Population	References
57.9	American (8th–10th centuries)	(Saunders 1978)
70	Indian (21st century)	(Saikia et al. 2017)
73.9	American (Eskimo, 16th–18th centuries)	(Saunders 1978)
75.6	French (9th–12th centuries)	(Verna 2014)
81.4	Polish (20th century)	(Polguj et al. 2011)
82.8	American (17th–18th centuries)	(Saunders 1978)
83.7	German (anatomical collection)	(Natsis et al. 2007)
84.3	French (19th–20th centuries)	(Verna 2014)

Table 1.14. *Frequency of suprascapular notching on dry bone*

1.2.3.4. *Association(s)*

A trend for bilaterality of AOV is observed.

1.2.4. *Unfused acromial epiphysis*

1.2.4.1. *Definition*

– Synonyms: Bipartite acromion and non-fusion of the acromion process.

– Latin translation: *Os acromiale.*

Description: There is lack of fusion of one of the three ossification centers and persistence of the immature form of the acromion after 20 years.

There are two hypotheses for its origin: one genetic and the other mechanical (intensive use of the rotator cuff muscles).

1.2.4.2. *Illustrations*

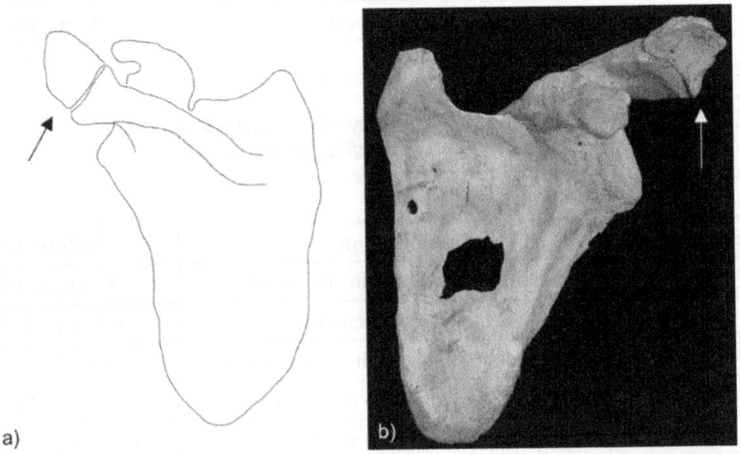

Figure 1.15. *Unfused acromial epiphysis: (a) diagram, left scapula, posterior view; (b) dry bone, left scapula, anterior view. For a color version of this figure, see www.iste.co.uk/verna/asymptomatic.zip*

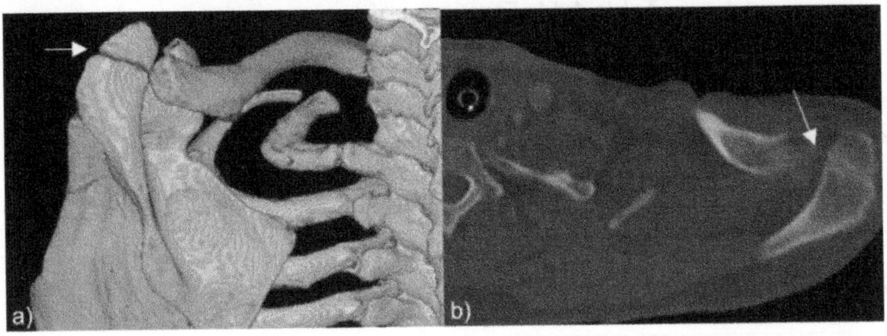

Figure 1.16. *(a) 3D reconstruction, left scapula, posterosuperior view; (b) DICOM image, left scapula, axial section. For a color version of this figure, see www.iste.co.uk/verna/asymptomatic.zip*

1.2.4.3. *Population frequencies*

A difference in frequency is observed between the different studies presented in the tables. It can be explained by a difference in behavior. The unfused acromial epiphysis is a marker of posture, and intensive use of the shoulder may prevent its fusion. Some populations may have had more intensive shoulder activity than others.

Frequency	Population	References
0.7	Korean	(Kumar et al. 2013)
2	Mediterranean	(Verna 2014)

Table 1.15. *Frequency of unfused acromial epiphysis in medical imaging*

Frequency	Population	References
3.6	American (17th–18th centuries)	(Saunders 1978)
3.9	American (8th–10th centuries)	(Saunders 1978)
5.6	French (19th–20th centuries)	(Verna 2014)
6	French (Neolithic)	(Thomas 2011)
6.7	French (4th–17th centuries)	(Debono 2006)
7.6	American (Eskimo, 16th–18th centuries)	(Saunders 1978)
7.7	Danish (11th–12th centuries)	(Case et al. 2005)
8	American (20th century)	(Sammarco 2000; Hunt and Bullen 2007)
11	French (9th–12th centuries)	(Verna 2014)
18.2	South African (20th century)	(Case et al. 2005)
20	French (8th–12th centuries)	(Crubézy 1988)

Table 1.16. *Frequency of unfused acromial epiphysis on dry bone*

1.2.4.4. *Association(s)*

Unfused acromial epiphysis does not show a relationship to sex except in the study by Hunt and Bullent (2007), showing a higher frequency in American women. Regarding laterality, Verna's (2014) study shows a trend toward unilaterality.

The left side is affected as much as the right, except in the study by Case et al. (2005) that showed a preference for the left side.

1.2.5. *Circumflex sulcus*

1.2.5.1. *Definition*

– Synonyms: None.

– Description: It is located on the posterolateral border in the form of a shallow, horizontal groove. It corresponds to the imprint left on the bone by the circumflex artery during growth.

1.2.5.2. *Illustrations*

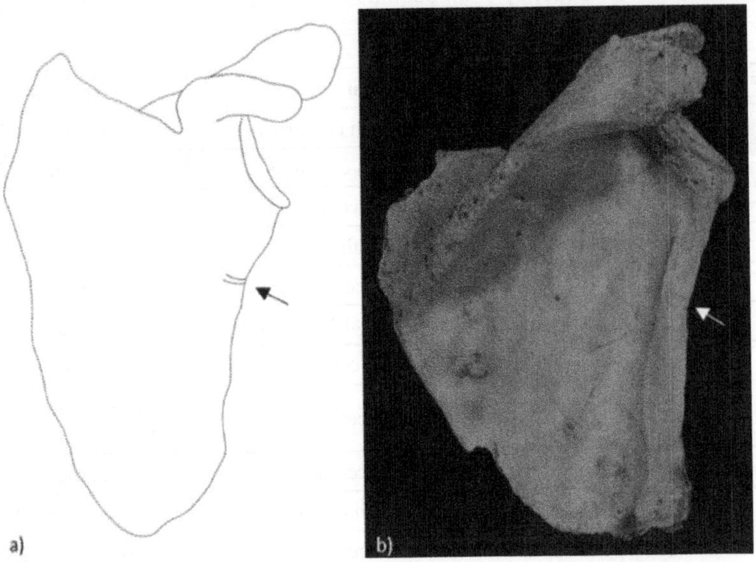

Figure 1.17. *Circumflex sulcus: (a) diagram, left scapula, anterior view; (b) dry bone, right scapula, posterior view. For a color version of this figure, see www.iste.co.uk/verna/asymptomatic.zip*

1.2.5.3. *Population frequencies*

A difference in frequency is visible between the studies. This can be explained in two ways: an AOV population dependent or a difference in rating.

Indeed, this AOV is the consequence of the imprint left by the circumflex artery on the bone and, depending on its depth (more or less marked), it becomes subjective at the time of rating and thus leads to a frequency difference.

Frequency	Population	References
	No study	

Table 1.17. *Frequency of circumflex sulcus in medical imaging*

Frequency	Population	References
23.8	French (19th–20th centuries)	(Verna 2014)
24.1	French (9th–12th centuries)	(Verna 2014)
73.7	American (17th–18th centuries)	(Saunders 1978)
77.7	American (Eskimo, 16th–18th centuries)	(Saunders 1978)
85.5	American (8th–10th centuries)	(Saunders 1978)
86.8	American (20th century)	(Gray 1942)

Table 1.18. *Frequency of circumflex sulcus on dry bone*

1.2.5.4. *Association(s)*

Saunders (1978) highlights an increase in the frequency of this AOV with age.

In his study, the sample of individuals consisted of both immature and adult individuals.

Immature individuals did not show this AOV due to unfinished skeletal maturation, so it seems normal to observe an increase in frequency contrary to other studies where only adult individuals were selected (Verna 2014).

The *circumflex sulcus* was more present on right scapulae (Saunders 1978).

1.2.6. *Acromial articular facet*

1.2.6.1. *Definition*

– Synonym: Humeral facet.

– Description: It is located on the anteroinferior surface of the acromial process due to the friction of the proximal extremity of the humerus on the lower surface of the acromion.

1.2.6.2. *Illustrations*

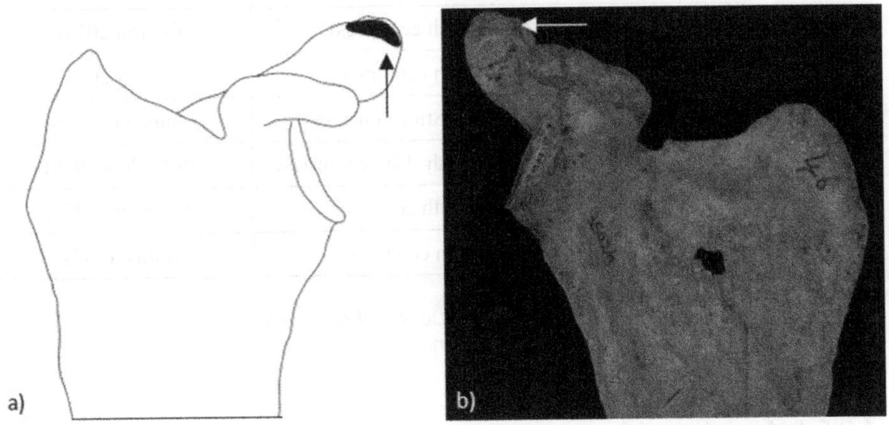

Figure 1.18. *Acromial articular facet: (a) diagram, left scapula, anterior view; (b) dry bone, right scapula, anterior view. For a color version of this figure, see www.iste.co.uk/verna/asymptomatic.zip*

1.2.6.3. *Population frequencies*

Frequency	Population	References
	No study	

Table 1.19. *Frequency of the acromial articular facet in medical imaging*

Frequency	Population	References
2	American (17th–18th centuries)	(Saunders 1978)
5.8	American (8th–10th centuries)	(Saunders 1978)
10.3	Scottish (19th century)	(Miles 1996)
11.8	American (Eskimo, 16th–18th centuries)	(Saunders 1978)
14.8	French (19th–20th centuries)	(Verna 2014)
22.1	American (20th century)	(Gray 1942)

Table 1.20. *Frequency of the acromial articular facet on dry bone*

1.2.6.4. *Association(s)*

No association has been found for this AOV.

On the other hand, this facet is quickly associated with pathological damage, since on this same site a facet with arthritic edges and the presence of porosities may replace the smooth acromial facet.

Therefore, care should be taken with this AOV and the rating should be adapted to the demographic profile of its study population.

1.3. Humerus

The humerus is formed from a primary ossification center constituting the diaphysis. Two epiphyses, one proximal and one distal, are composed of several secondary ossification centers.

Fusion of the epiphyses to the diaphysis varies between males and females, but occurs between 11 and 20 years of age (Scheuer and Black 2004; Scheuer et al. 2009).

A total of four AOVs are presented for the humerus and all are observable from a very young age.

1.3.1. *Septal aperture*

1.3.1.1. *Definition*

– Synonyms: Its synonyms are perforation of the coronoid-olecranon and supratrochlear foramen.

– Description: It involves perforation of the cortical bone separating the olecranon and the coronoid fossa.

1.3.1.2. *Illustrations*

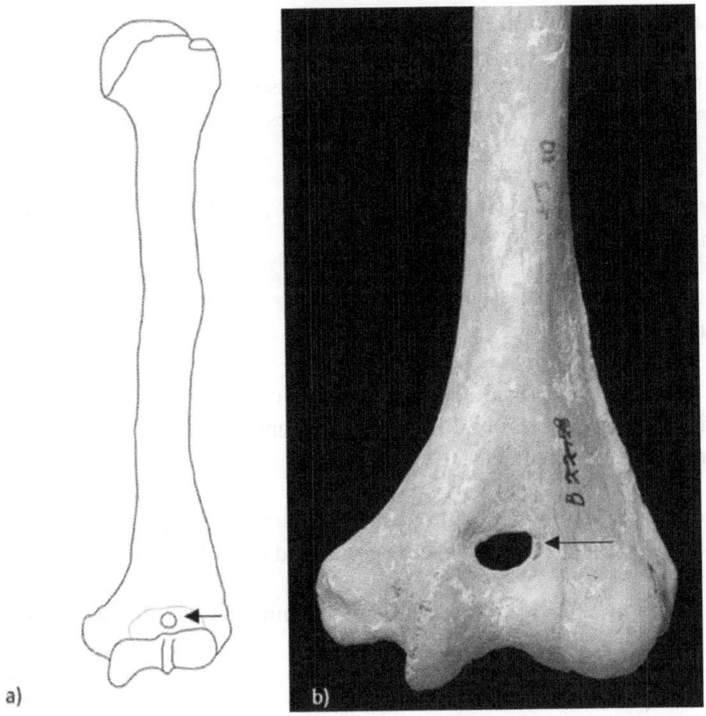

Figure 1.19. *Septal aperture: (a) diagram, left humerus, anterior view; (b) dry bone, left humerus, anterior view. For a color version of this figure, see www.iste.co.uk/verna/asymptomatic.zip*

1.3.1.3. *Population frequencies*

This AOV shows a great variability of frequency within a population. It appears a lot in anthropological studies (it is easy to observe), but seems to be of very little interest in view of its great variability, with sometimes significant frequencies, unless one can relate the frequency to the historical period.

Frequency	Population	References
8.6	Turkish	(Koyun et al. 2011)

Table 1.21. *Frequency of septal aperture in medical imaging*

Frequency	Population	References
6.2	French (5th–7th centuries)	(Castex 1994)
9.4	French (4th–17th centuries)	(Debono 2006)
12.6	French (19th–20th centuries)	(Verna 2014)
15	French (4th–6th centuries)	(Castex 1994)
18.7	French (8th–12th centuries)	(Crubézy 1988)
18.8	Japanese (20th century)	(Akabori 1934)
19	French (5th–6th centuries)	(Castex 1994)
20	French (1st–5th centuries)	(Murail 1996)
21	French (Neolithic)	(Thomas 2011)
21.4	French (9th–12th centuries)	(Verna 2014)
25	French (12th century)	(Castex 1994)
29.8	French (4th–5th centuries)	(Castex 1994)
33	American (Eskimo, 16th–18th centuries)	(Saunders 1978)
34.3	Indian (20th century)	(Nayak et al. 2009)
38.3	American (17th–18th centuries)	(Saunders 1978)
45.9	American (8th–10th centuries)	(Saunders 1978)

Table 1.22. *Frequency of septal aperture on dry bone*

1.3.1.4. *Association(s)*

Septal aperture is more prevalent in women and on the left side (Castex 1994; Debono 2006; Koyun 2011; Thomas 2011; Verna 2014).

This link to sex can be explained by a hypothesis putting forward a link between bone robustness and the presence of AOV: the more slender the bone, the higher the frequency of AOV (Benfer and McKern 1966).

On the other hand, regarding bilaterality, the preference changes across studies. Murail (1996) and Verna (2014) show a preference for unilaterality, in contrast to Koyun et al. (2011) where AOV is predominantly bilateral (Turkish population).

1.3.2. *Supracondyloid process*

1.3.2.1. *Definition*

– Synonym: Supratrochlear spur.

– Description: It has a small prominent bone located on the anteromedial surface in the distal third at a variable distance from the medial epicondyle.

1.3.2.2. *Illustrations*

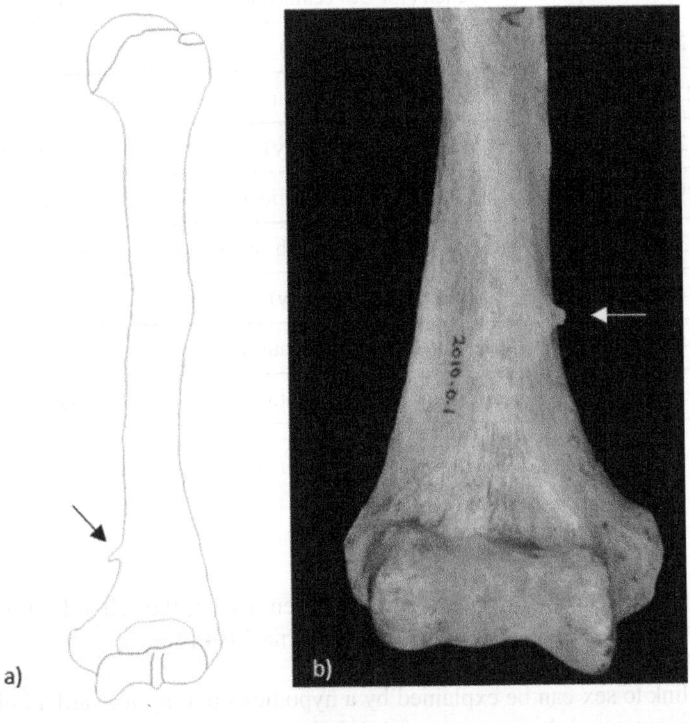

Figure 1.20. *Supracondyloid process: (a) diagram, left humerus, anterior view; (b) dry bone, right humerus, anterior view. For a color version of this figure, see www.iste.co.uk/verna/asymptomatic.zip*

1.3.2.3. *Population frequencies*

Frequency	Population	References
1	Turkish	(Aydinlioglu et al. 2010)

Table 1.23. *Frequency of the supracondyloid process in medical imaging*

Frequency	Population	References
0	American (8th–10th centuries)	(Saunders 1978)
0.3	Indian (20th century)	(Kumar and Mehta 2008)
0.6	French (9th–12th centuries)	(Verna 2014)
0.9	American (Eskimo, 16th–18th centuries)	(Saunders 1978)
0.9	American (17th–18th centuries)	(Saunders 1978)
1.1	French (4th–5th centuries)	(Castex 1994)
1.3	German (20th century)	(Natsis 2008)
1.5	French (1st–5th centuries)	(Murail 1996)
1.6	French (4th–17th centuries)	(Debono 2006)
3	French (Neolithic)	(Thomas 2011)
4.1	French (19th–20th centuries)	(Verna 2014)

Table 1.24. *Frequency of the supracondyloid process on dry bone*

1.3.2.4. *Association(s)*

This AOV is mostly observed unilaterally.

1.3.3. *Supracondyloid tunnel*

1.3.3.1. *Definition*

– Synonyms: None.

– Description: It is located on the medial border of the humerus, 2–3 cm above the epicondyle. It can be used for the passage of the anterior and posterior ulnar arteries.

1.3.3.2. *Illustrations*

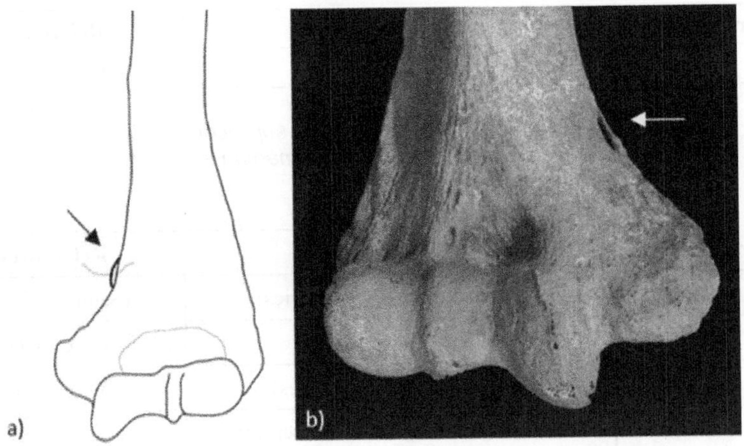

Figure 1.21. *Supracondyloid tunnel: (a) diagram, left humerus, anterior view; (b) dry bone, right humerus, anterior view. For a color version of this figure, see www.iste.co.uk/verna/asymptomatic.zip*

1.3.3.3. *Population frequencies*

Frequency	Population	References
No study		

Table 1.25. *Frequency of the supracondyloid tunnel in medical imaging*

Frequency	Population	References
0	French (19th–20th centuries)	(Verna 2014)
1.7	French (4th–17th centuries)	(Debono 2006)
1.7	French (9th–12th centuries)	(Verna 2014)

Table 1.26. *Frequency of the supracondyloid tunnel on dry bone*

1.3.3.4. *Association(s)*

No association was found for this AOV.

1.3.4. *Translucency of the septum*

1.3.4.1. *Definition*

– Synonyms: None.

– Description: The bony lamina separating the olecranon fossa from the coronoid fossa is thin enough to cause translucency visible to the naked eye.

This is not an intermediate stage of septal aperture.

1.3.4.2. *Illustrations*

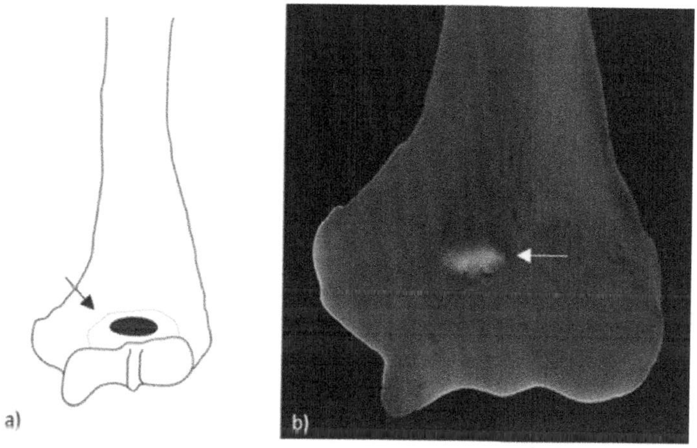

Figure 1.22. *Translucency of the septum: (a) diagram, left humerus, anterior view; (b) dry bone, left humerus, anterior view. For a color version of this figure, see www.iste.co.uk/verna/asymptomatic.zip*

1.3.4.3. *Population frequencies*

A significant difference in frequency is observed between the three samples presented in Table 1.28. One hypothesis can be put forward as to this large difference in frequency: the collection dating from the 20th century presents relatively "fresh" bones, still fatty, and the olecranon fossa is thin and often shows translucency.

The bones of the collection dating from the 9th to 12th centuries are denser, with sometimes concretions of sediment making this area more opaque and thus being able to explain this difference of frequency.

Frequency	Population	Reference(s)
	No study	

Table 1.27. *Frequency of translucency of the septum in medical imaging*

Frequency	Population	Reference
7.6	French (9th–12th centuries)	(Verna 2014)
38.5	French (4th–17th centuries)	(Debono 2006)
63.5	French (19th–20th centuries)	(Verna 2014)

Table 1.28. *Frequency of translucency of the septum on dry bone*

1.3.4.4. *Association(s)*

Translucency of the septum was found preferentially in females in the study by Debono (2006). For all the samples, bilaterality was the majority.

1.4. Radius

The ossification of the radius starts from a primary ossification center composing the diaphysis. The proximal and distal epiphyses each have a secondary ossification center.

A secondary center may also be present at the tuberosity. Fusion is completed around 17–20 years of age (Scheuer and Black 2004; Scheuer et al. 2009).

Two AOVs can be observed on the radius.

1.4.1. *Radial tuberosity orientation*

1.4.1.1. *Definition*

– Synonyms: None.

– Description: Radial tuberosity can have three different orientations in relation to the interbody ridge: medial (position 1), anteromedial (position 2) and anterior (position 3).

The orientation can be related to the biomechanical activity of the elbow.

1.4.1.2. *Illustrations*

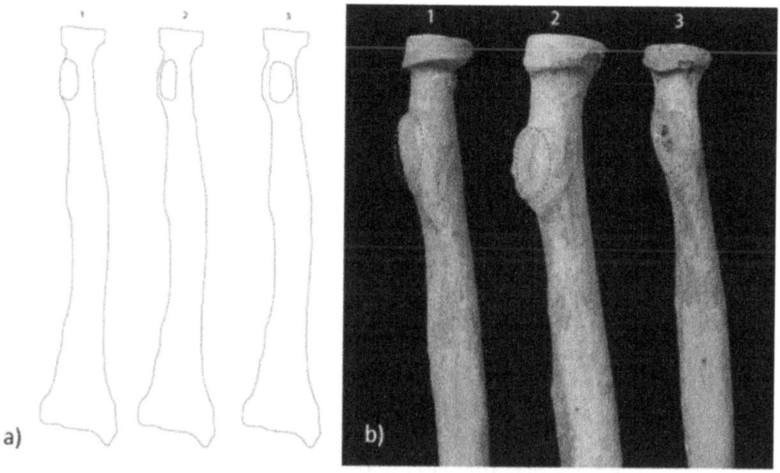

Figure 1.23. *Radial tuberosity orientation: (a) diagram, right radius, anterior views; (b) dry bone, right radius, anterior views. For a color version of this figure, see www.iste.co.uk/verna/asymptomatic.zip*

1.4.1.3. *Population frequencies*

Only position 3 has a similar frequency to the literature, as it is the most easily identifiable position by its medial orientation. The other two positions are relatively close and therefore more difficult to distinguish clearly. Perhaps this difficulty in sizing, as well as the significant intra-/inter-observer errors, explains this difference in frequency between studies within the same population (French population only).

Frequency	Population	References
No study		

Table 1.29. *Frequency of radial tuberosity orientation in medical imaging*

	Frequency	Population	References
Position 1	77.2–85.4	French (19th–20th centuries)	(Verna 2014)
	69.6–77.8	French (9th–12th centuries)	(Verna 2014)
	59.1	French (4th–17th centuries)	(Debono 2006)
Position 2	11.7–21.8	French (19th–20th centuries)	(Verna 2014)
	18.1–24.6	French (9th–12th centuries)	(Verna 2014)
	38.7	French (4th–17th centuries)	(Debono 2006)
Position 3	1–3	French (19th–20th centuries)	(Verna 2014)
	4.1–5.8	French (9th–12th centuries)	(Verna 2014)
	2.2	French (4th–17th centuries)	(Debono 2006)

Table 1.30. *Frequency of radial tuberosity orientation on dry bone according to the position of the tuberosity*

1.4.1.4. *Association(s)*

Regarding laterality, position 1 is predominantly bilateral in Verna's (2014) study, which is not found in Debono's (2006) study. On the other hand, a preference for the left side (Debono 2006) is found, contrary to Verna's (2014) study.

For position 2, the opposite is observed for both bilaterality and direction (Debono 2006).

Position 3 is present on both the left and the right (Debono 2006).

1.4.2. *Double distal radial articular facet*

1.4.2.1. *Definition*

– Synonyms: None.

– Description: The inferior articular surface of the distal end of the radius may present a double surface separated by a bony bulge.

1.4.2.2. *Illustrations*

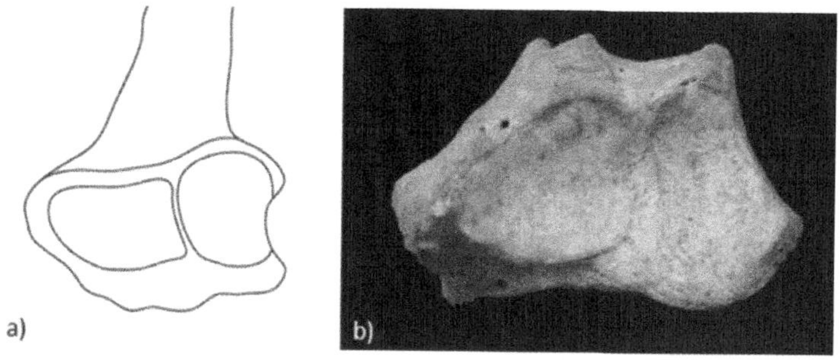

Figure 1.24. *Double distal radial articular facet: (a) diagram, right radius, inferior view; (b) dry bone, left radius, inferior view. For a color version of this figure, see www.iste.co.uk/verna/asymptomatic.zip*

1.4.2.3. *Population frequencies*

Frequency	Population	References
No study		

Table 1.31. *Frequency of the double distal radial articular facet in medical imaging*

Frequency	Population	References
1	French (19th–20th centuries)	(Verna 2014)
4	French (9th–12th centuries)	(Verna 2014)

Table 1.32. *Froquoncy of tho doublo distal radial articular facet on dry bone*

1.4.2.4. *Association(s)*

No association was found for this AOV.

1.5. Ulna

The ulna is formed from a primary ossification center giving the diaphysis. The proximal epiphysis ossifies from two secondary centers, while the distal epiphysis ossifies from a single secondary center. The epiphyses fuse to the diaphysis between 12 and 20 years of age, with the proximal fusing first (Scheuer and Black 2004;

Scheuer et al. 2009). The three AOVs shown can be observed from an early age on the ulna.

1.5.1. *Medial trochlear notch*

1.5.1.1. *Definition*

– Synonyms: None.

– Description: The articular surface of the ulnar head may have a single notch on the medial side.

1.5.1.2. *Illustrations*

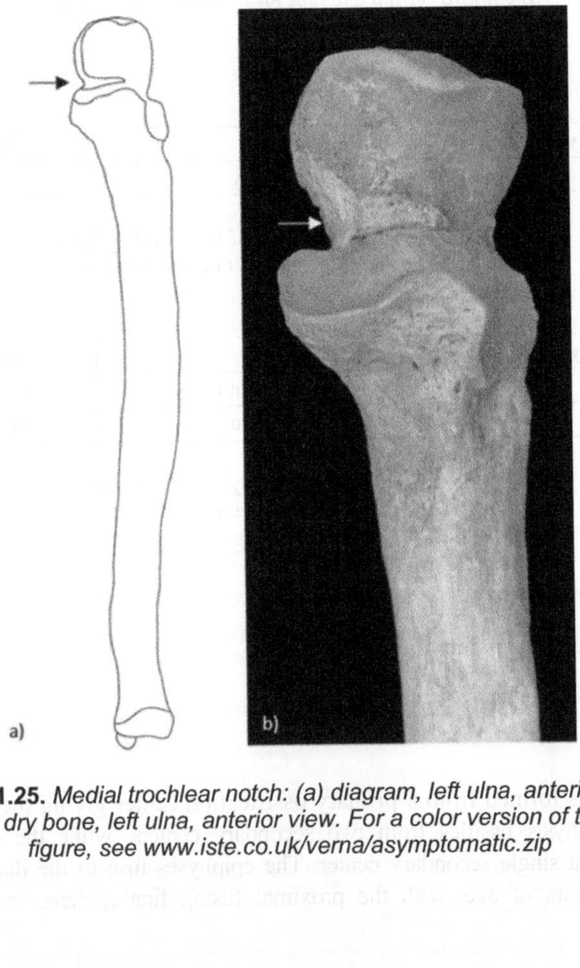

Figure 1.25. *Medial trochlear notch: (a) diagram, left ulna, anterior view; (b) dry bone, left ulna, anterior view. For a color version of this figure, see www.iste.co.uk/verna/asymptomatic.zip*

1.5.1.3. *Population frequencies*

Frequency	Population	References
No study		

Table 1.33. *Frequency of the medial trochlear notch in medical imaging*

Frequency	Population	References
40.2	French (9th–12th centuries)	(Verna 2014)
73.5	French (19th–20th centuries)	(Verna 2014)

Table 1.34. *Frequency of the medial trochlear notch on dry bone*

1.5.1.4. *Association(s)*

No association was found for this AOV.

1.5.2. *Lateral trochlear notch*

1.5.2.1. *Definition*

– Synonyms: None.

– Description: The articular surface of the ulnar head may have a single notch on the lateral side.

1.5.2.2. *Illustrations*

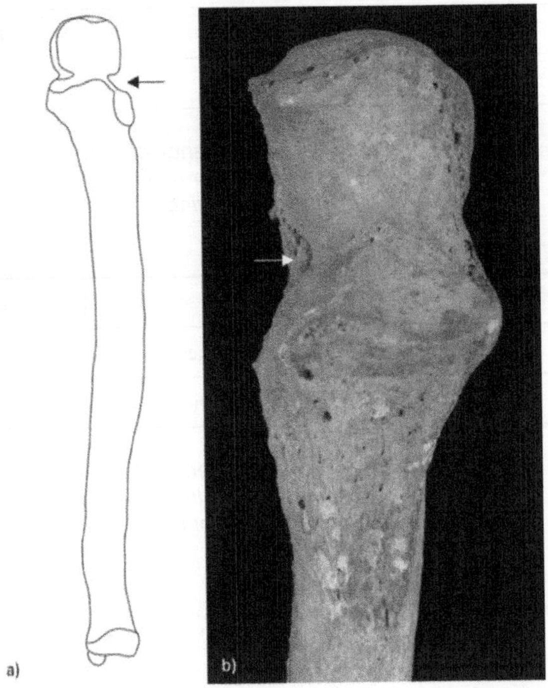

Figure 1.26. *Lateral trochlear notch: (a) diagram, left ulna, anterior view; (b) dry bone, right ulna, anterior view. For a color version of this figure, see www.iste.co.uk/verna/asymptomatic.zip*

1.5.2.3. *Population frequencies*

Frequency	Population	References
	No study	

Table 1.35. *Frequency of the lateral trochlear notch in medical imaging*

Frequency	Population	References
12.3	French (9th–12th centuries)	(Verna 2014)
33.3	French (19th–20th centuries)	(Verna 2014)

Table 1.36. *Frequency of the lateral trochlear notch on dry bone*

1.5.2.4. *Association(s)*

No association was found for this AOV.

1.5.3. *Trochlear groove*

1.5.3.1. *Definition*

– Synonyms: None.

– Description: The trochlear surface may have two distinct facets with the presence of a groove.

1.5.3.2. *Illustrations*

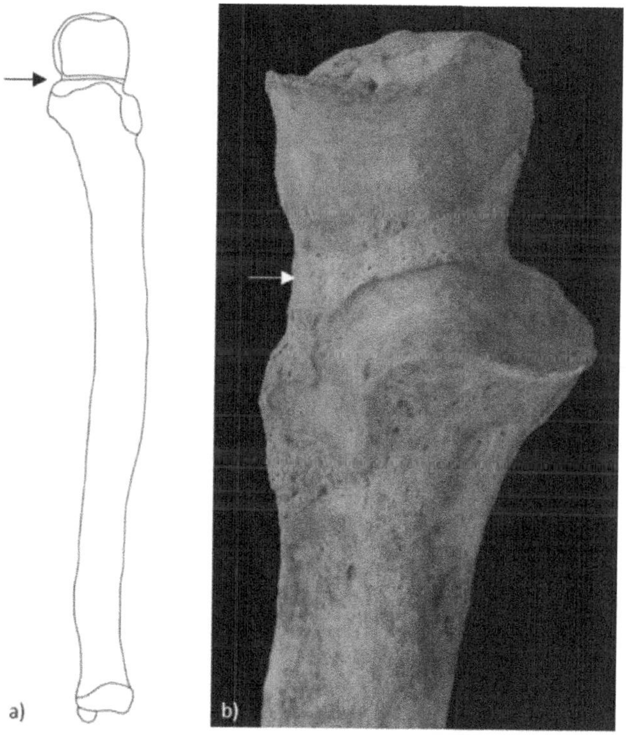

Figure 1.27. *Trochlear groove: (a) diagram, left ulna, anterior view; (b) dry bone, right ulna, anterior view. For a color version of this figure, see www.iste.co.uk/verna/asymptomatic.zip*

1.5.3.3. *Population frequencies*

The trochlear groove has a frequency ranging from 5.1 to 44%. It has a large variability in frequency with no particular explanation.

Frequency	Population	References
	No study	

Table 1.37. *Frequency of the trochlear groove in medical imaging*

Frequency	Population	References
5.1	French (5th–7th centuries)	(Castex 1994)
5.4	French (21st century)	(Castex 1994)
6.1	French (4th–5th centuries)	(Castex 1994)
11.5	American (8th–10th centuries)	(Saunders 1978)
12.3	French (19th–20th centuries)	(Verna 2014)
27.3	American (Eskimo, 16th–18th centuries)	(Saunders 1978)
33.1	French (9th–12th centuries)	(Verna 2014)
38	American (17th–18th centuries)	(Saunders 1978)
44.1	French (4th–17th centuries)	(Debono 2006)

Table 1.38. *Frequency of the trochlear groove on dry bone*

1.5.3.4. *Association(s)*

Trochlear groove here is predominantly present in males in Verna's (2014) study, unlike Debono's (2006) study showing a preference for females.

The frequency seems to increase with age (Castex 1994; Debono 2006).

Bilaterality dominated for the study by Verna (2014). In contrast, no link with side was found (Debono 2006).

2

The Thorax

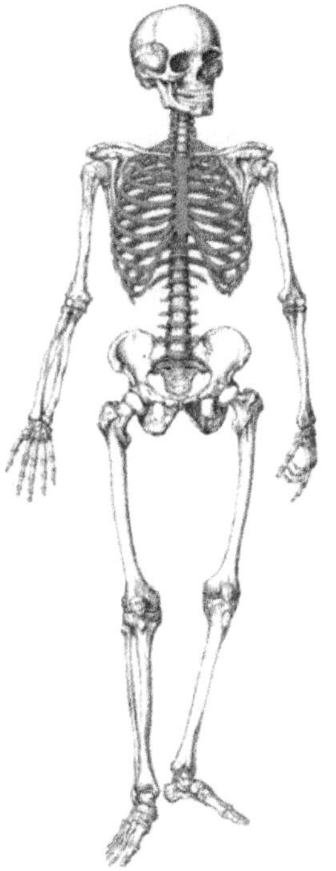

Figure 2.1. *Human skeleton. The thorax is presented in red. For a color version of this figure, see www.iste.co.uk/verna/asymptomatic.zip*

This chapter focuses on the AOVs present within the thorax. This anatomical region has a large number of variations, especially on the sternum. The sternum is relatively fragile and is less well preserved as dry bone, which may limit the study and interpretation of data for this anatomical region in archeological studies in particular.

2.1. Sternum

The ossification centers of the sternum appear from the fifth to sixth fetal months. They begin with the manubrium followed by those of the sternal body. For the manubrium, most of the time, only one ossification center is present, but it is possible to have two. A fusion defect between these two centers can be possible and lead to a foramen, but this is a very rare case. The ossification centers of the sternal body vary from one to several per sternum. The sternebrae fuse along the midline to form four sternebrae, which form the immature sternum. Failure to fuse along the midline results in a sternal cleft that may be partial or complete. The fusion of the sternae with each other starts at the lower part of the sternum, that is, between the third and the fourth, and then goes up to reach the sternal body. Failure of the sternae to fuse together can lead to the formation of a foramen. The last to fuse to the mesosternum is the first sternebrae, between 15 and 20 years of age. The ossification center of the xiphoid process appears later, between 3 and 6 years of age, and may begin to ossify (Scheuer and Black 2004; Scheuer et al. 2009).

On the sternum, nine AOVs are observable, including the conformation of the xiphoid process, which can take four forms (absent, single, double or triple). Hypostotic AOVs are only seen from the age of 20 years onwards when skeletal maturation is complete.

2.1.1. *Sternal cleft*

2.1.1.1. *Definition*

– Synonyms: None.

– Description: It includes fusion defect of the midline of the sternum.

There are three forms:

– superior (upper part of the sternal body and/or manubrium) (1, 2);

– inferior (lower part of the sternal body and/or xiphoid process) (3);

– complete (or also called bifid sternum, the rarest form) (4).

2.1.1.2. *Illustrations*

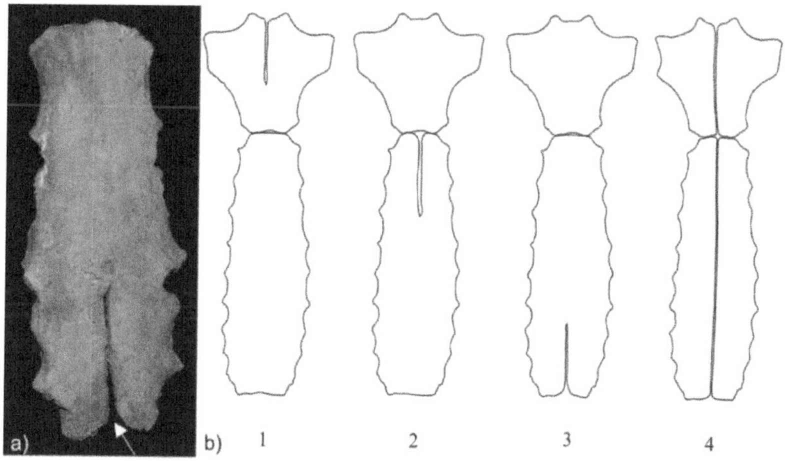

Figure 2.2. *Sternal cleft: (a) dry bone, sternal body, anterior view; (b) diagram, four cleft shapes, anterior views. For a color version of this figure, see www.iste.co.uk/verna/asymptomatic.zip*

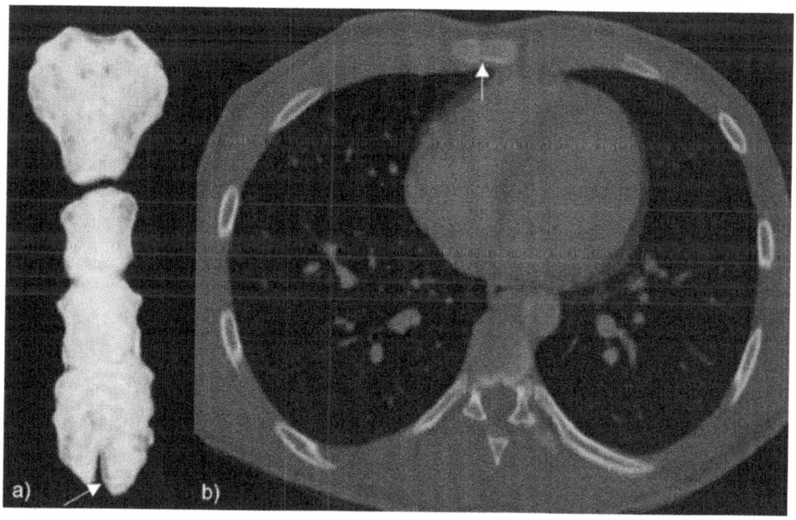

Figure 2.3. *(a) 3D reconstruction, anterior view; (b) DICOM image, axial section. For a color version of this figure, see www.iste.co.uk/verna/asymptomatic.zip*

2.1.1.3. *Population frequencies*

Frequency	Population	References
0.4	Mediterranean	(Verna 2014)
1.4	Turkish	(Yekeler et al. 2006)
4	Turkish	(Bayarogullari et al. 2014)

Table 2.1. *Frequency of sternal clefts (all forms combined) in medical imaging*

Frequency	Population	References
0	French (9th–12th centuries)	(Verna 2014)

Table 2.2. *Frequency of sternal clefts (all forms combined) on dry bone*

A study conducted on Turkish individuals under 25 years of age shows a significant frequency for this AOV (4%) (Bayarogullari et al. 2014).

Therefore, attention must be paid to the age of the study population to avoid introducing ossification bias.

2.1.1.4. *Association(s)*

No association was found for this AOV.

2.1.2. *Non-fusion of sternebrae*

2.1.2.1. *Definition*

– Synonyms: None.

– Description: It includes lack of fusion of one or more sternebrae to the rest of the sternal body.

2.1.2.2. *Illustrations*

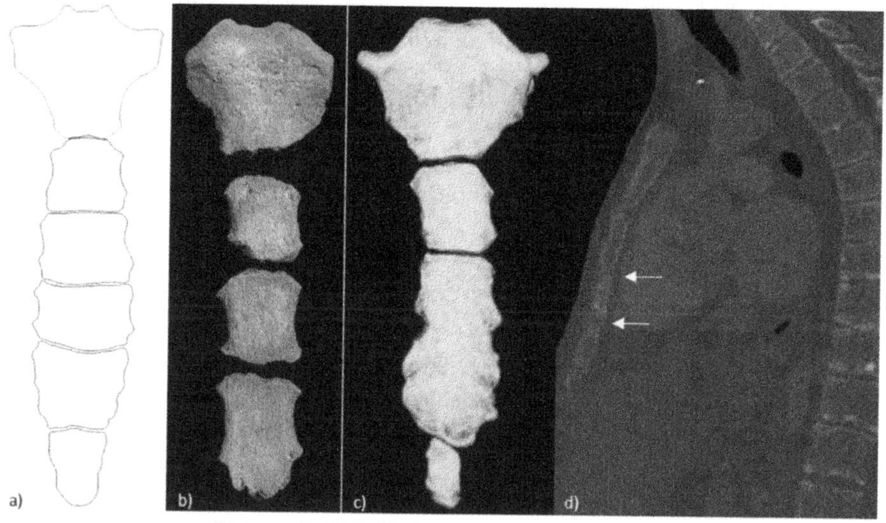

Figure 2.4. *Non-fusion of sternebrae: (a) diagram, anterior view; (b) dry bone, anterior view; (c) 3D reconstruction, anterior view; (d) DICOM image, sagittal section. For a color version of this figure, see www.iste.co.uk/verna/asymptomatic.zip*

2.1.2.3. *Population frequencies*

Frequency	Population	References
1.8	Mediterranean	(Verna 2014)
2.4	Turkish	(Bayarogullari et al. 2014)

Table 2.3. *Frequency of non-fusion of sternebrae in medical imaging*

Frequency	Population	References
0.6	French (5th–7th centuries)	(Castex 1994)
1.5	French (9th–12th centuries)	(Verna 2014)

Table 2.4. *Frequency of non-fusion of sternebrae on dry bone*

2.1.2.4. *Association(s)*

This AOV is predominantly found in males in the study by Bayarogullari et al. (2014), which was not observed in the study by Verna (2014). On the other hand, it predominantly affects 20- to 29-year-olds; a phenomenon of fusion of the sternae to the rest of the body with age can be envisaged, hence a peak before 30 years of age. This AOV can only be considered a variation from 20 years of age onwards once the ossification of the sternum is complete; otherwise, it is still the immature form of this bone.

2.1.3. *Misplaced manubrium-mesosternal joint*

2.1.3.1. *Definition*

– Synonyms: None.

– Description: It includes fusion of the manubrium with the first sternum of the sternal body. The manubrium-body articulation is then made at the level of the third rib.

2.1.3.2. *Illustrations*

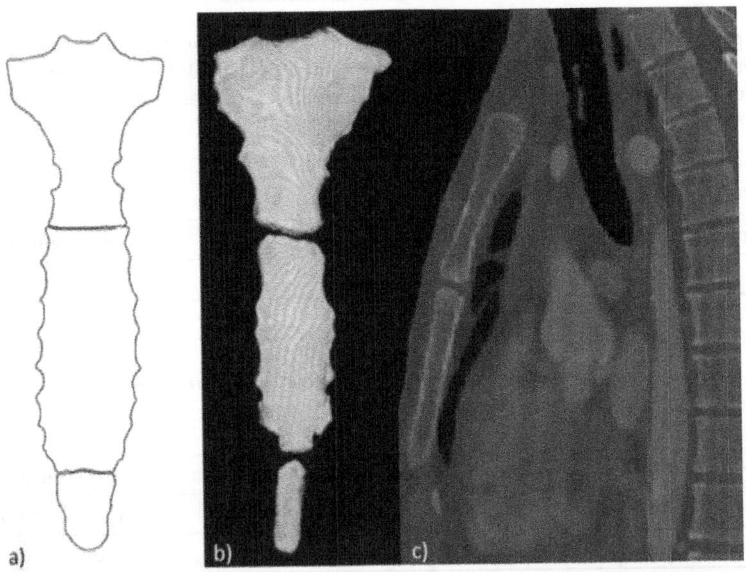

Figure 2.5. *Misplaced manubrium-mesosternal joint: (a) diagram, anterior view; (b) 3D reconstruction, anterior view; (c) DICOM image, sagittal section. For a color version of this figure, see www.iste.co.uk/verna/asymptomatic.zip*

2.1.3.3. *Population frequencies*

Frequency	Population	References
1.3	Mediterranean	(Verna 2014)

Table 2.5. *Frequency of misplaced manubrium-mesosternal joints in medical imaging*

Frequency	Population	References
0	French (9th–12th centuries)	(Verna 2014)

Table 2.6. *Frequency of misplaced manubrium-mesosternal joints on dry bone*

Very few works on AOVs study the misplaced manubrium-mesosternal joint. It is difficult to compare data for the latter.

2.1.3.4. *Association(s)*

No association was found for this AOV.

2.1.4. *Sternal foramen*

2.1.4.1. *Definition*

Synonyms: Sternal aperture, sternal perforation, midline sternal foramina and perforatio sterni.

– Latin translation: *Perforatio sterni.*

– Description: It includes an ossification defect usually occurring in the third or fourth segment at the time of fusion of the sternebrae.

2.1.4.2. *Illustrations*

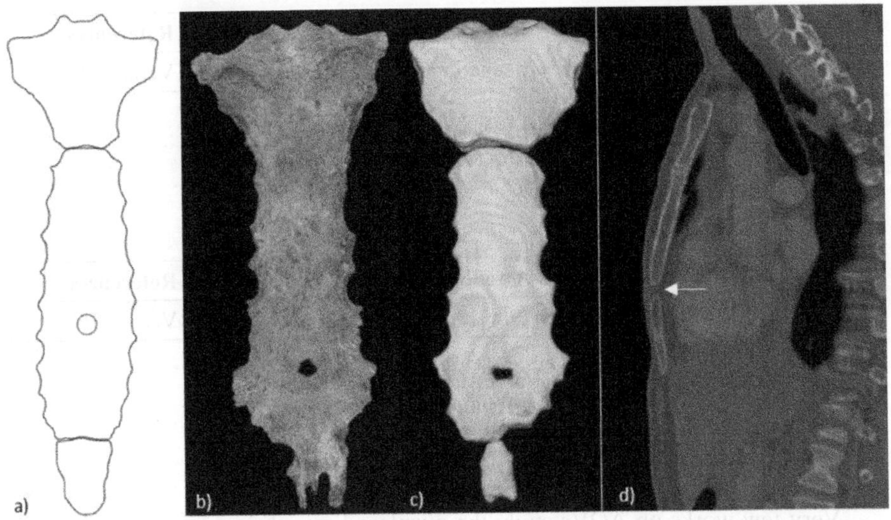

Figure 2.6. *Sternal foramen: (a) diagram, anterior view; (b) dry bone, anterior view; (c) 3D reconstruction, anterior view; (d) DICOM image, sagittal section. For a color version of this figure, see www.iste.co.uk/verna/asymptomatic.zip*

2.1.4.3. *Population frequencies*

Frequency	Population	References
3.3	Spanish	(Macaluso and Lucena 2014)
4.5	Turkish	(Yekeler et al. 2006)
4.8	Mediterranean	(Verna 2014)
5	Turkish	(Aktan and Savas 1998)
6	Turkish	(Bayarogullari et al. 2014)
6.7	American	(Cooper et al. 1988)
10.5	Brazilian woman	(Babinski et al. 2015)

Table 2.7. *Frequency of sternal foramen in medical imaging*

Frequency	Population	References
1.1	American (8th–10th centuries)	(Saunders 1978)
2.1	French (5th–7th centuries)	(Castex 1994)
2.4	American (Eskimo, 16th–18th centuries)	(Saunders 1978)
5	Slovak (11th–12th centuries)	(Masnicova and Benus 2003)
5	Slovak (9th century)	(Masnicova and Benus 2003)
5.2	French (9th–12th centuries)	(Verna 2014)
5.3	American (17th–18th centuries)	(Saunders 1978)
5.5	French (4th–17th centuries)	(Debono 2006)
12	Indian (20th century)	(Singh and Pathak 2013)

Table 2.8. *Frequency of sternal foramen on dry bone*

The sternal foramen is the most studied AOV of the sternum. The average frequency is 5% in all populations.

2.1.4.4. *Association(s)*

This AOV is observed more in men, which is confirmed by the literature (Masnicova and Benus 2003; Bayarogullari et al. 2014).

2.1.5. *Manubriosternal fusion*

2.1.5.1. *Definition*

– Synonyms: None.

– Description: It includes fusion between the manubrium and the sternal body.

2.1.5.2. *Illustrations*

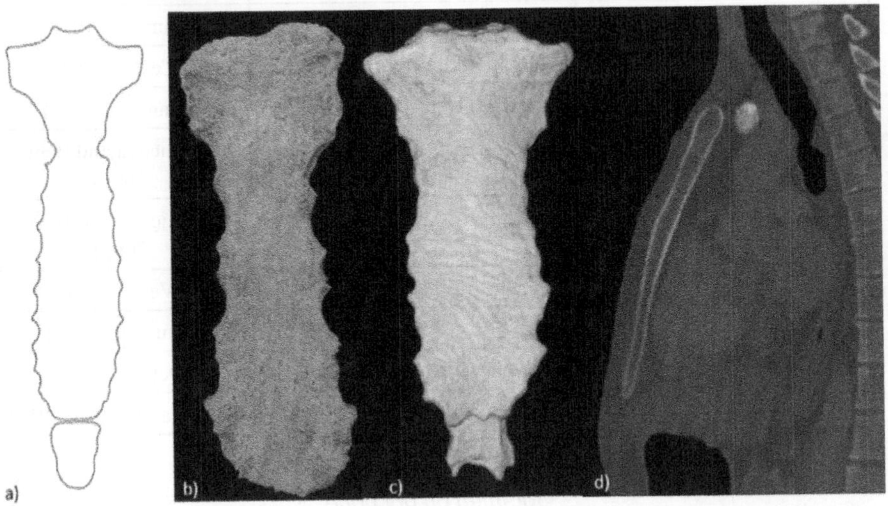

Figure 2.7. *Manubriosternal fusion: (a) diagram, anterior view; (b) dry bone, anterior view; (c) 3D reconstruction, anterior view; (d) DICOM image, sagittal section. For a color version of this figure, see www.iste.co.uk/verna/asymptomatic.zip*

2.1.5.3. *Population frequencies*

Frequency	Population	References
15.7	Mediterranean	(Verna 2014)
19.6	Turkish	(Yekeler et al. 2006)
22.4	Turkish	(Turkay et al. 2017)

Table 2.9. *Frequency of manubriosternal fusion in medical imaging*

Frequency	Population	References
11.6	French (9th–12th centuries)	(Verna 2014)
23.6	Indian (20th century)	(Singh and Pathak 2013)

Table 2.10. *Frequency of manubriosternal fusion on dry bone*

2.1.5.4. *Association(s)*

Manubriosternal fusion is not related to sex. Moreover, those over 30 years of age are more affected; late fusion of the manubrium and the body is considered. A study performed on individuals under 25 years of age shows that no fusion is visible before 16 (Bayarogullari et al. 2014). Therefore, attention must be paid to the age of individuals to avoid over-/underestimating the frequency of this AOV.

2.1.6. *Manubrium foramen*

2.1.6.1. *Definition*

– Synonyms: None.

– Description: It occurs due to a fusion defect of the two ossification centers of the manubrium.

2.1.6.2. *Illustrations*

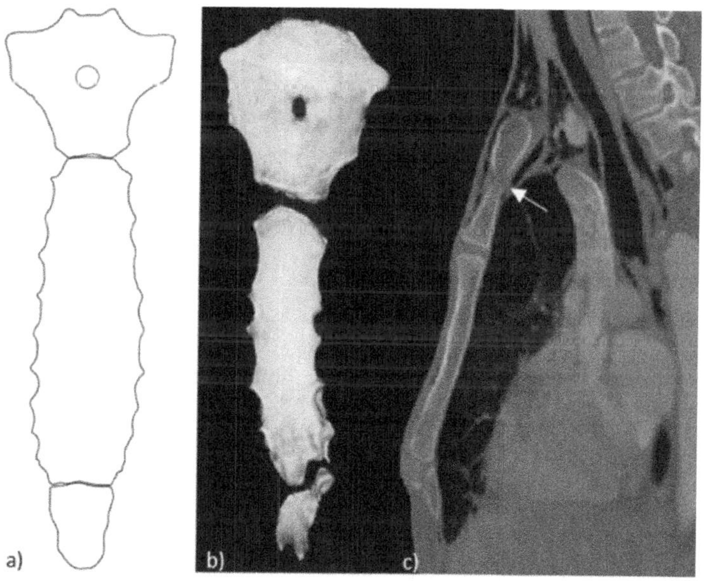

Figure 2.8. *Manubrium foramen: (a) diagram, anterior view; (b) 3D reconstruction, anterior view; (c) DICOM image, sagittal section. For a color version of this figure, see www.iste.co.uk/verna/asymptomatic.zip*

2.1.6.3. *Population frequencies*

Only two cases have been observed in the literature for this variation. It is very rare.

Frequency	Population	References
0.1	Mediterranean	(Verna 2014)
0.4	American	(Cooper et al. 1988)

Table 2.11. *Frequency of manubrium foramen in medical imaging*

Frequency	Population	References
0	French (9th–12th centuries)	(Verna 2014)

Table 2.12. *Frequency of manubrium foramen on dry bone*

2.1.6.4. *Association(s)*

No association was found for this AOV.

2.1.7. *Xiphoid foramen*

2.1.7.1. *Definition*

– Synonyms: None.

– Description: It occurs due to an ossification defect (same principle as for the sternal foramen).

2.1.7.2. *Illustrations*

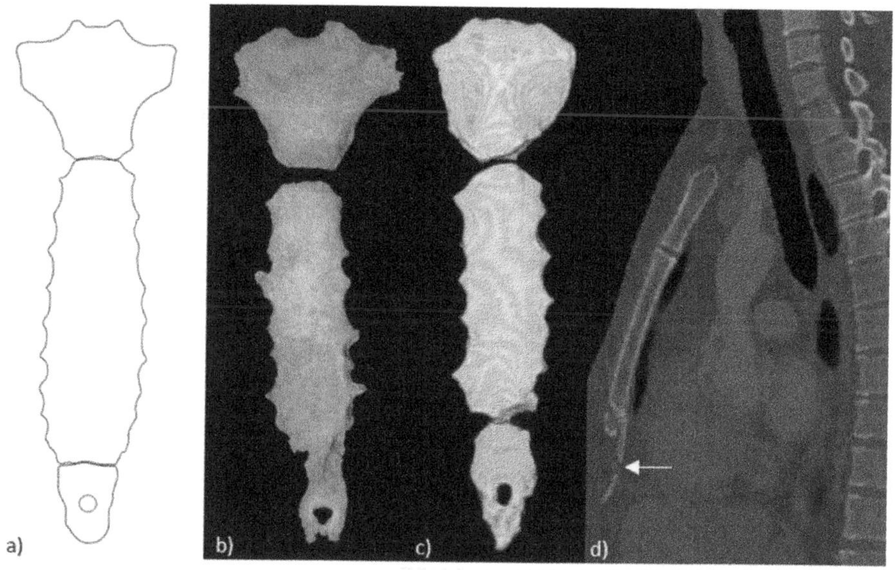

Figure 2.9. *Xiphoid foramen: (a) diagram, anterior view; (b) dry bone, anterior view; (c) 3D reconstruction, anterior view; (d) DICOM image, sagittal section. For a color version of this figure, see www.iste.co.uk/verna/asymptomatic.zip*

2.1.7.3. *Population frequencies*

Frequency	Population	References
9.6	Mediterranean	(Verna 2014)
27.4	Turkish	(Yekeler et al. 2006)
43.2	Turkish	(Akin et al. 2011)

Table 2.13. *Frequency of xiphoid foramen in medical imaging*

Frequency	Population	References
6.7	French (9th–12th centuries)	(Verna 2014)

Table 2.14. *Frequency of xiphoid foramen on dry bone*

2.1.7.4. *Association(s)*

This AOV is more often observed in men.

Regarding age, it is mostly found in people over 50 years old. For this variation, it is also necessary to pay attention to the age of the sample (individual or multiple) in order not to bias the frequency.

Individuals must have completed their skeletal maturation to observe this AOV.

2.1.8. *Single xiphoid process*

2.1.8.1. *Definition*

– Synonyms: None.

– Description: The end of the xiphoid process is simple (one extremity only).

2.1.8.2. *Illustrations*

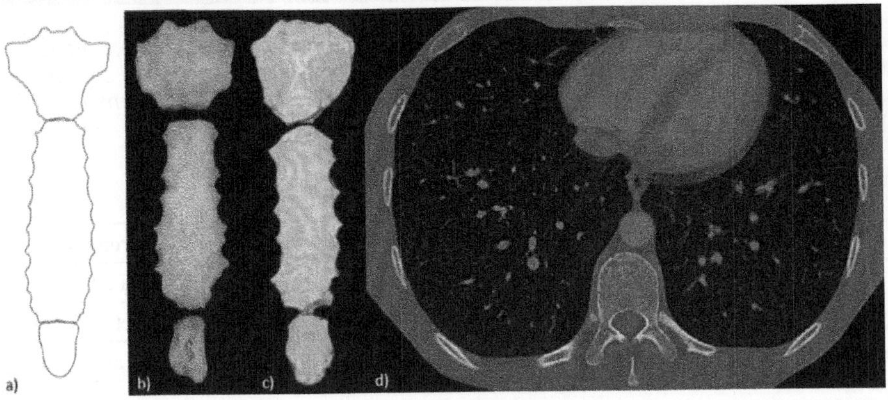

Figure 2.10. *Single xiphoid process: (a) diagram, anterior view; (b) dry bone, anterior view; (c) 3D reconstruction, anterior view; (d) DICOM image, axial section. For a color version of this figure, see www.iste.co.uk/verna/asymptomatic.zip*

2.1.8.3. *Population frequencies*

Frequency	Population	References
40.4	Turkish	(Bayarogullari et al. 2014)
54.9	Spanish	(Macaluso et al. 2014)
56.2	Mediterranean	(Verna 2014)
62.6	Turkish	(Akin et al. 2011)
71	Turkish	(Yekeler et al. 2006)

Table 2.15. *Frequency of the single xiphoid process in medical imaging*

Frequency	Population	References
	No study	

Table 2.16. *Frequency of the single xiphoid process on dry bone*

2.1.8.4. *Association(s)*

No association was found for this AOV.

2.1.9. *Bifurcated xiphoid process*

2.1.9.1. *Definition*

– Synonyms: None.

– Description: The end of the xiphoid process has two extremities.

2.1.9.2. *Illustrations*

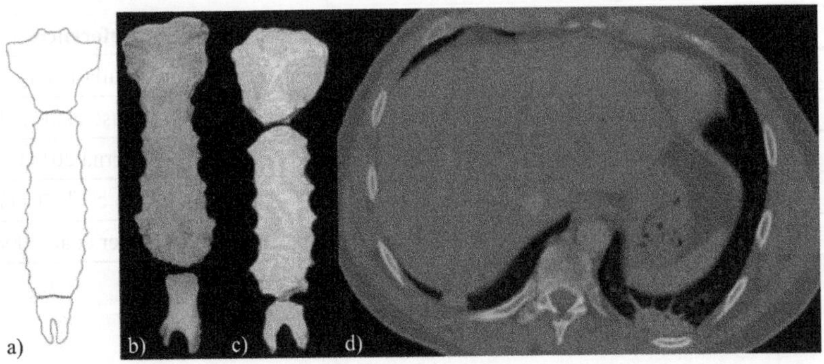

Figure 2.11. *Bifurcated xiphoid process: (a) diagram, anterior view; (b) dry bone, anterior view; (c) 3D reconstruction, anterior view; (d) DICOM image, axial section. For a color version of this figure, see www.iste.co.uk/verna/asymptomatic.zip*

2.1.9.3. *Population frequencies*

Frequency	Population	References
27.2	Turkish	(Yekeler et al. 2006)
31.2	Mediterranean	(Verna 2014)
32.8	Turkish	(Akin et al. 2011)
41.6	Spanish	(Macaluso et al. 2014)

Table 2.17. *Frequency of the bifurcated xiphoid process in medical imaging*

Frequency	Population	References
14.9	Indian (20th century)	(Singh 2013)

Table 2.18. *Frequency of the bifurcated xiphoid process on dry bone*

2.1.9.4. *Association(s)*

This conformation is mostly found in people over 30 years old.

2.1.10. *Triple xiphoid process*

2.1.10.1. *Definition*

– Synonyms: None.

– Description: The end of the xiphoid process has three distinct extremities.

2.1.10.2. *Illustrations*

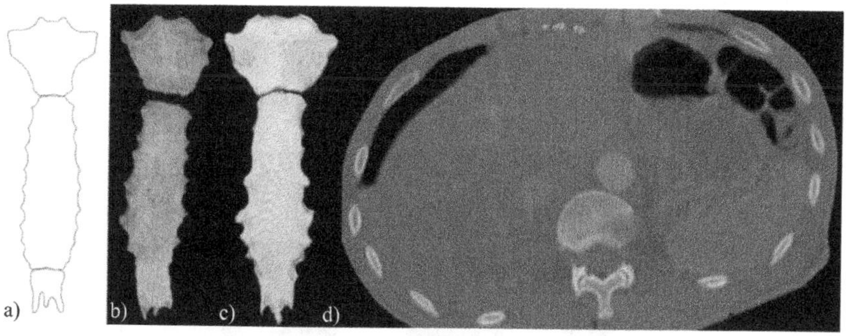

Figure 2.12. *Triple xiphoid process: (a) diagram, anterior view; (b) dry bone, anterior view; (c) 3D reconstruction, anterior view; (d) DICOM image, axial section. For a color version of this figure, see www.iste.co.uk/verna/asymptomatic.zip*

2.1.10.3. *Population frequencies*

Frequency	Population	References
0.7	Turkish	(Yekeler et al. 2006)
2.1	Mediterranean	(Verna 2014)
4.6	Turkish	(Akin et al. 2011)
16	Spanish	(Macaluso et al. 2014)

Table 2.19. *Frequency of the triple xiphoid process in medical imaging*

Frequency	Population	References
No study		

Table 2.20. *Frequency of the triple xiphoid process on dry bone*

2.1.10.4. *Association(s)*

This conformation is mostly found in people over 30 years old.

2.1.11. *Absence of the xiphoid process*

2.1.11.1. *Definition*

– Synonyms: None.

– Description: The xiphoid process is not present and there is no ossification.

2.1.11.2. *Illustrations*

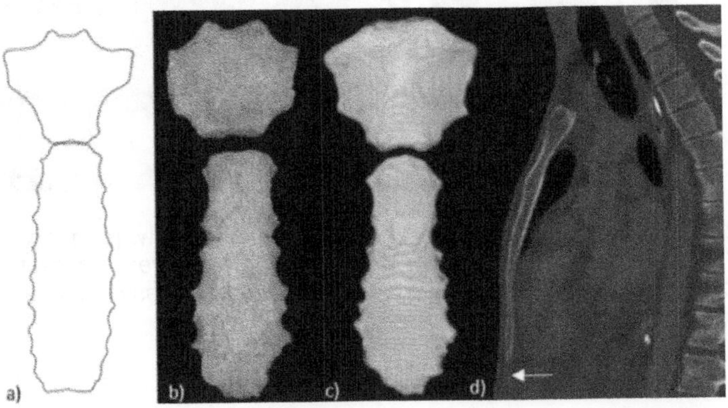

Figure 2.13. *Absence of the xiphoid process: (a) diagram, anterior view; (b) dry bone, anterior view; (c) 3D reconstruction, anterior view; (d) DICOM image, sagittal section. For a color version of this figure, see www.iste.co.uk/verna/asymptomatic.zip*

2.1.11.3. *Population frequencies*

Frequency	Population	References
1.1	Turkish	(Yekeler et al. 2006)
6.6	Spanish	(Macaluso et al. 2014)
10.5	Mediterranean	(Verna 2014)

Table 2.21. *Frequency of absence of the xiphoid process in medical imaging*

Frequency	Population	References
9	Indian (20th century)	(Singh and Pathak 2013)

Table 2.22. *Frequency of absence of the xiphoid process on dry bone*

2.1.11.4. *Association(s)*

Females have a higher frequency than males for this AOV. The absence of the xiphoid process is predominantly present in those under 25 years of age.

2.1.12. *Suprasternal bones*

2.1.12.1. *Definition*

– Synonyms: None.

– Description: It includes small bones located on the upper edge of the manubrium. One or two bones may be present.

2.1.12.2. *Illustrations*

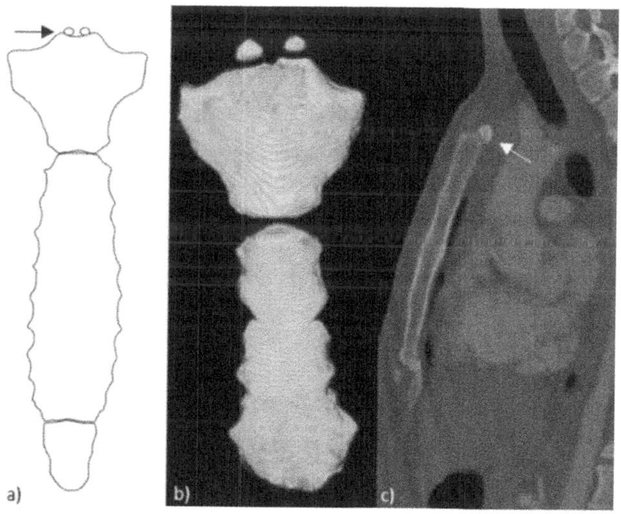

Figure 2.14. *Suprasternal bone: (a) diagram, anterior view; (b) 3D reconstruction, anterior view; (c) DICOM image, sagittal section. For a color version of this figure, see www.iste.co.uk/verna/asymptomatic.zip*

2.1.12.3. *Population frequencies*

Frequency	Population	References
2	Turkish	(Bayarogullari et al. 2014)
3	Turkish	(Turkay et al. 2017)
3.3	Mediterranean	(Verna 2014)
4.1	Turkish	(Yekeler et al. 2006)

Table 2.23. *Frequency of suprasternal bones in medical imaging*

Frequency	Population	References
No study		

Table 2.24. *Frequency of suprasternal bones on dry bone*

2.1.12.4. *Association(s)*

Suprasternal bones are mostly found in individuals over 20 years of age.

2.2. Ribs

Ribs form from a primary ossification center and three secondary centers. The vertebral extremity fuses to the rib body last, around 17–25 (Scheuer and Black 2004; Scheuer et al. 2009).

Seven AOVs are present on the ribs and can be observed from an early age.

2.2.1. *Cervical rib*

2.2.1.1. *Definition*

– Synonyms: None.

– Latin translation: *Costa cervicalis*.

– Description: It is located at the level of the seventh cervical vertebra, either in the form of rudimentary structures or in the form of a whole rib with a joint on the manubrium.

This is due to cranial displacement of the cervicothoracic border (Barnes 2012).

2.2.1.2. *Illustrations*

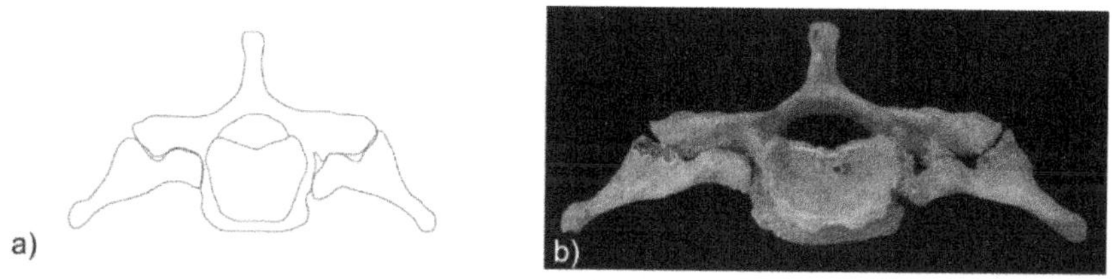

Figure 2.15. *Cervical rib on the seventh cervical vertebra: (a) diagram, seventh cervical, superior view; (b) dry bone, seventh cervical, superior view. For a color version of this figure, see www.iste.co.uk/verna/asymptomatic.zip*

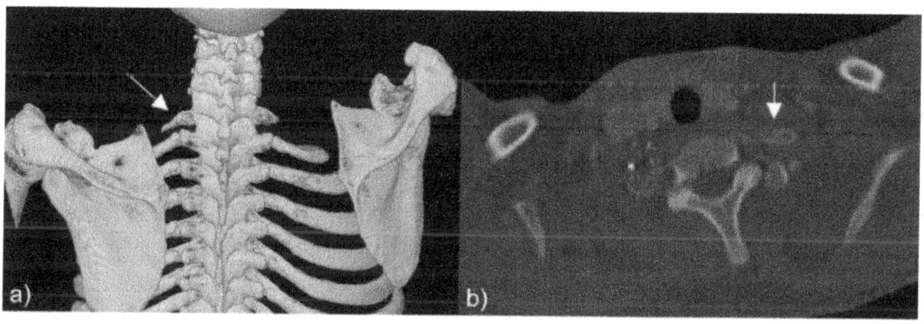

Figure 2.16. *Cervical rib on the seventh cervical vertebra: (a) 3D reconstruction, posterior view; (b) DICOM image, axial section. For a color version of this figure, see www.iste.co.uk/verna/asymptomatic.zip*

2.2.1.3. *Population frequencies*

The highest frequency was found for dry bone samples. The possible confusion with the long transverse processes of the seventh cervical vertebra that can mimic a cervical rib may explain this difference in frequency between the two media (imaging and dry bone). Is it easier to differentiate a cervical rib from a long transverse process on dry bone than on imaging and thus not underestimate the frequency? These long processes have a frequency ranging from 2–5%. This can therefore be a hypothesis.

Frequency	Population	References
0.3	Japanese	(Takeuchi et al. 2016)
0.7	English	(Brewin et al. 2009)
2	American	(Viertel et al. 2012)
2.1	Mediterranean	(Verna 2014)
3.6	Turkish	(Davran et al. 2017)

Table 2.25. *Frequency of the cervical rib in medical imaging*

Frequency	Population	References
0.3	American	(Karapetian 2017)
4.5	French (9th–12th centuries)	(Verna 2014)
5.8	French (19th–20th centuries)	(Verna 2014)

Table 2.26. *Frequency of the cervical rib on dry bone*

2.2.1.4. *Association(s)*

The cervical rib is found more often in women than in men and preferentially on the left.

2.2.2. *Lumbar rib*

2.2.2.1. *Definition*

– Synonyms: None.

– Latin translation: *Costa lumbalis*.

– Description: It is located at the level of the first lumbar vertebra.

This is due to a caudal shift in the thoracolumbar junction (Barnes 2012).

2.2.2.2. *Illustrations*

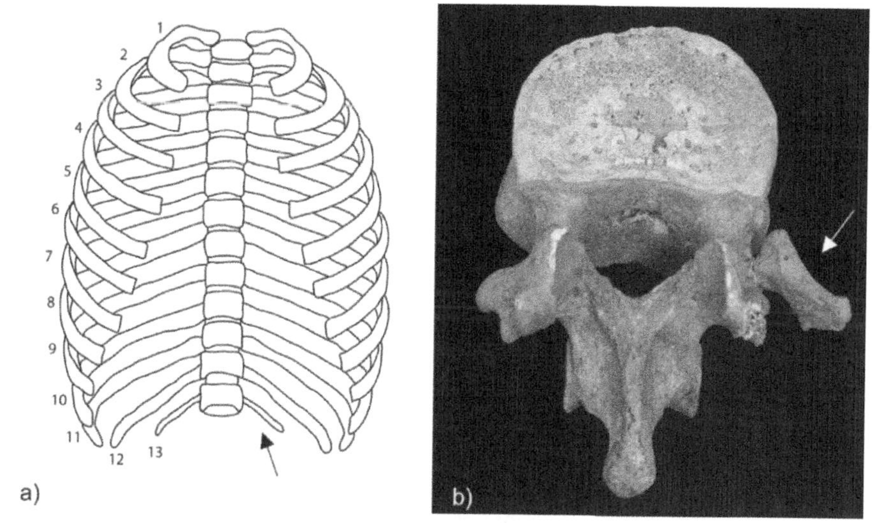

Figure 2.17. *Lumbar rib: (a) diagram, anterior view; (b) dry bone, first lumbar, posterior-superior view. For a color version of this figure, see www.iste.co.uk/verna/asymptomatic.zip*

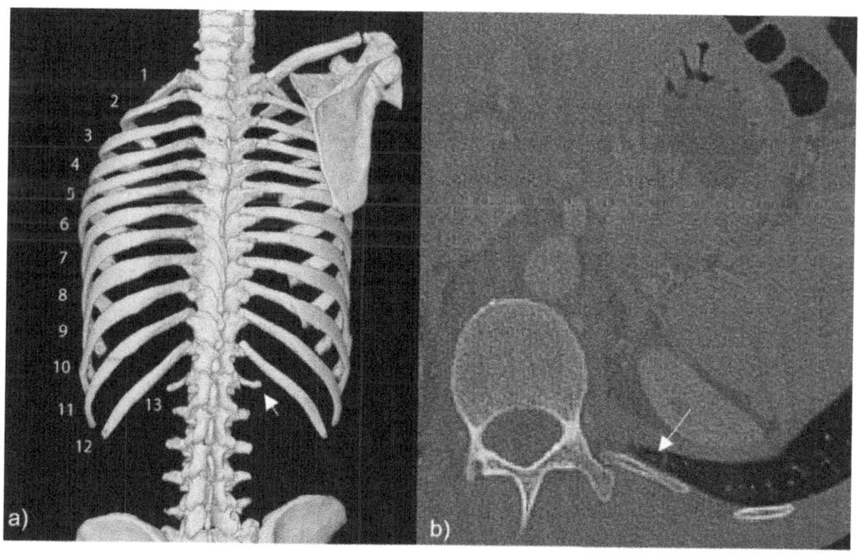

Figure 2.18. *Lumbar rib: (a) 3D reconstruction, posterior view; (b) DICOM image, axial section. For a color version of this figure, see www.iste.co.uk/verna/asymptomatic.zip*

2.2.2.3. *Population frequencies*

Frequency	Population	References
1.3	Turkish	(Davran et al. 2017)
4.8	Mediterranean	(Verna 2014)
5.8	Japanese	(Nakajima et al. 2014)

Table 2.27. *Frequency of lumbar ribs in medical imaging*

Frequency	Population	References
2.6	French (9th–12th centuries)	(Verna 2014)
3	Slovak (11th–12th centuries)	(Masnicova and Benus 2003)

Table 2.28. *Frequency of lumbar ribs on dry bone*

2.2.2.4. *Association(s)*

It is more often found in men and bilaterally.

2.2.3. ***Bifid rib***

2.2.3.1. *Definition*

– Synonyms: None.

– Description: It includes sternal extremity with two distinct joints.

2.2.3.2. *Illustrations*

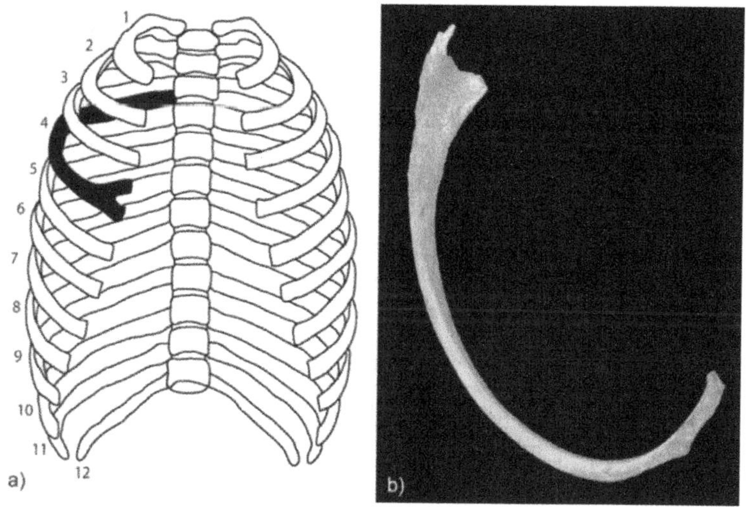

Figure 2.19. *Bifid rib: (a) diagram, anterior view; (b) dry bone, right rib, inferior view. For a color version of this figure, see www.iste.co.uk/verna/asymptomatic.zip*

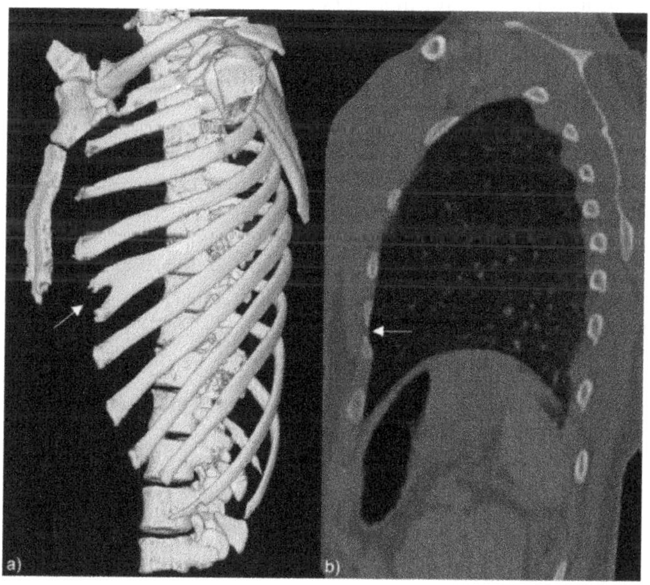

Figure 2.20. *Bifid rib: (a) 3D reconstruction, left rib, left lateral view; (b) DICOM image, left rib, sagittal section. For a color version of this figure, see www.iste.co.uk/verna/asymptomatic.zip*

2.2.3.3. *Population frequencies*

Frequency	Population	References
Case report	Korean	(Song et al. 2009)
1.2	Korean	(Lim et al. 1982)
1.3	Mediterranean	(Verna 2014)
3.3	Spanish	(Macaluso et al. 2014)

Table 2.29. *Frequency of bifid ribs in medical imaging*

Frequency	Population	References
1.5	French (9th–12th centuries)	(Verna 2014)

Table 2.30. *Frequency of bifid ribs on dry bone*

2.2.3.4. *Association(s)*

No association was found for this AOV.

2.2.4. *Costal fusion*

2.2.4.1. *Definition*

– Synonym: Merged ribs.

– Description: It includes fusion of two ribs at the body level. Be careful to distinguish a congenital fusion from a fusion due to trauma (consolidation of a rib fracture) (Barnes 2012).

2.2.4.2. *Illustrations*

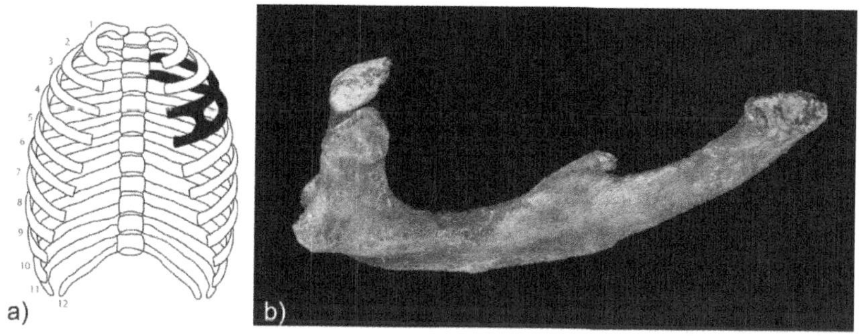

Figure 2.21. *Costal fusion: (a) diagram, anterior view; (b) dry bone, right ribs, superior view. For a color version of this figure, see www.iste.co.uk/verna/asymptomatic.zip*

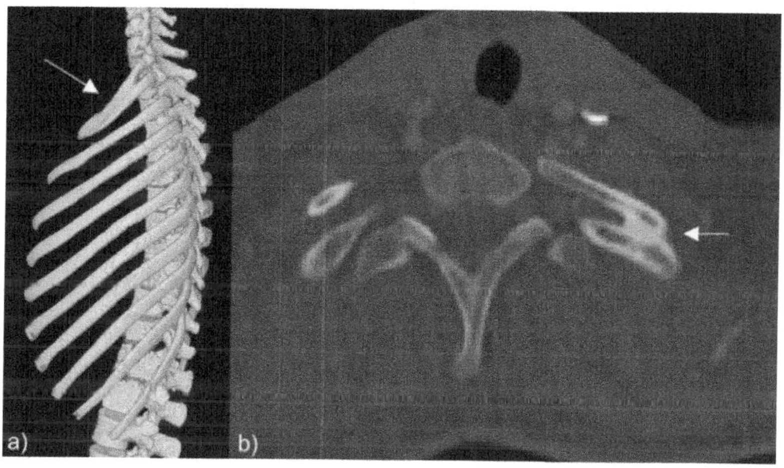

Figure 2.22. *Costal fusion (a) 3D reconstruction, left ribs, left lateral view; (b) DICOM image, left ribs, axial section. For a color version of this figure, see www.iste.co.uk/verna/asymptomatic.zip*

2.2.4.3. *Population frequencies*

Frequency	Population	References
0.3	Korean	(Lim et al. 1982)
0.4	Mediterranean	(Verna 2014)

Table 2.31. *Frequency of costal fusion in medical imaging*

Frequency	Population	References
0.7	French (9th–12th centuries)	(Verna 2014)

Table 2.32. *Frequency of costal fusion on dry bone*

2.2.4.4. *Association(s)*

No association was found for this AOV.

2.2.5. *Sacral rib*

2.2.5.1. *Definition*

– Synonyms: None.

– Description: It is located at the level of the sacrum and hinges on a sacral vertebra.

We need to be careful to differentiate a congenital form from an acquired form (due to trauma) (Rashid et al. 2008).

2.2.5.2. *Illustration*

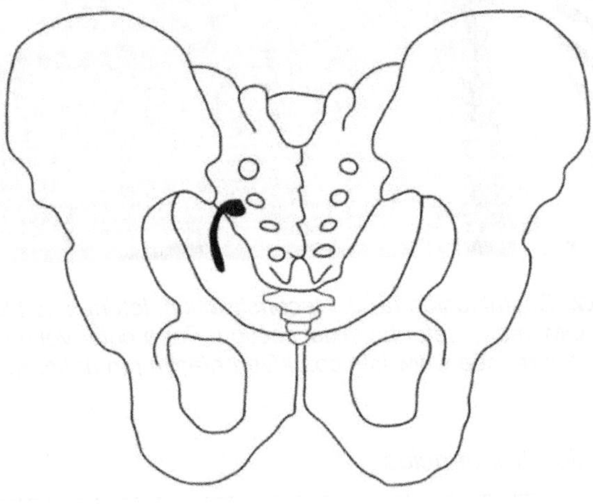

Figure 2.23. *Diagram, posterior view of the pelvis*

2.2.5.3. *Population frequencies*

Frequency	Population	References
0	Mediterranean	(Verna 2014)
Case report	Indian	(Rashid et al. 2008)
Case report	Japanese	(Miyakoshi et al. 2015)

Table 2.33. *Frequency of sacral ribs in medical imaging*

Frequency	Population	References
No study		

Table 2.34. *Frequency of sacral ribs on dry bone*

2.2.5.4. *Association(s)*

No association was found for this AOV.

2.2.6. *Intrathoracic rib*

2.2.6.1. *Definition*

– Synonyms: None.

– Description: It lies across the rib cage, perpendicular to other ribs and within the rib cage (Apaydin et al. 2009) or along a rib, parallel (Barnes 2012).

2.2.6.2. *Illustration*

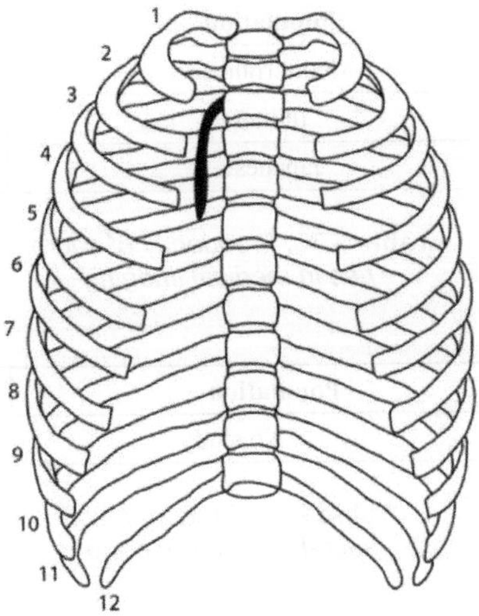

Figure 2.24. *Diagram, anterior view*

2.2.6.3. *Population frequencies*

Frequency	Population	References
0	Mediterranean	(Verna 2014)
Case report	Turkish	(Apaydin et al. 2009)

Table 2.35. *Frequency of intrathoracic ribs in medical imaging*

Frequency	Population	References
No study		

Table 2.36. *Frequency of intrathoracic ribs on dry bone*

2.2.6.4. *Association(s)*

No association was found for this AOV.

2.2.7. *Absence of the 12th rib*

2.2.7.1. *Definition*

Synonyms: None.

Description: The 12th rib is absent (unilaterally or bilaterally).

2.2.7.2. *Illustrations*

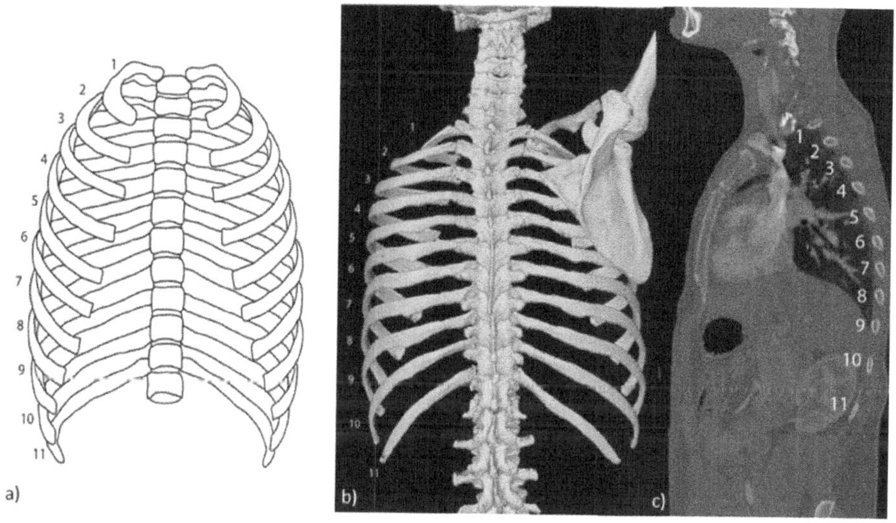

Figure 2.25. *Absence of the 12th rib: (a) diagram, anterior view; (b) 3D reconstruction, posterior view; (c) DICOM image, sagittal section. For a color version of this figure, see www.iste.co.uk/verna/asymptomatic.zip*

2.2.7.3. *Population frequencies*

Frequency	Population	References
0.9	Japanese	(Nakajima et al. 2014)
3.4	Mediterranean	(Verna 2014)

Table 2.37. *Frequency of absence of the 12th rib in medical imaging*

Frequency	Population	References
No study		

Table 2.38. *Frequency of absence of the 12th rib on dry bone*

2.2.7.4. *Association(s)*

It is mostly present in women and bilaterally.

3

The Vertebral Column

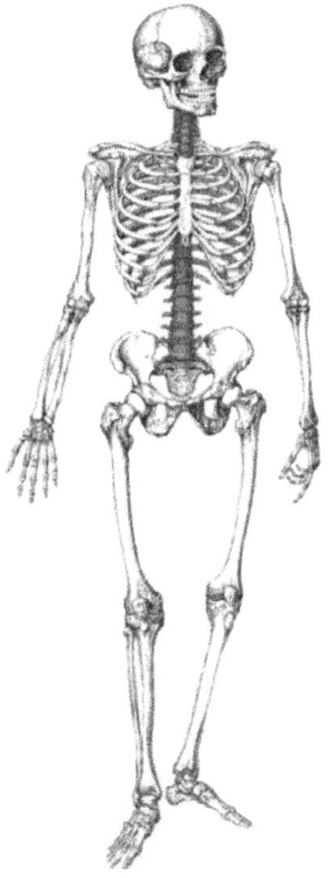

Figure 3.1. *Human skeleton. The vertebral column is presented in red. For a color version of this figure, see www.iste.co.uk/verna/asymptomatic.zip*

This chapter focuses on all the AOVs present within the vertebral column, from the first cervical vertebra to the fifth lumbar vertebra. The sacrum is presented in Chapter 4 (pelvic girdle). The coccyx has not been included in this book.

3.1. First cervical vertebra (atlas)

The first cervical vertebra is composed of three primary ossification centers. The two ossification centers of the posterior arch fuse at about 4–5 years of age. Failure to fuse results in *spina bifida occulta* or absence of the arch (in its entirety or only a part). The anterior arch is composed of a center that fuses to the rest of the vertebra at about 3–4 years of age. The transverse foramina are incomplete at birth and close by 3–4 years old. These foramina may never close (incomplete transverse foramina). The anterior arch may fuse in different ways, and variations may account for the absence of the anterior arch or the non-fusion of the arch (Scheuer and Black 2004; Scheuer et al. 2009).

Eight AOVs are present on the atlas. Hypostotic AOVs such as *spina bifida occulta* or incomplete transverse foramen are only observable from the age of about 10 years onwards so that the vertebra is mature and the AOV is not confused with an immature stage of the vertebra.

3.1.1. *Posterior atlas bridge*

3.1.1.1. *Definition*

– Synonym: Little posterior bridge.

– Latin translation: *Ponticulus posticus*.

– Description: It includes small bony bridge between the posterior portion of the superior articular process and the postero-lateral portion of the superior border of the posterior arch. This is due to the ossification of the lateral segment of the posterior atlanto-occipital ligament (Schilling et al. 2010).

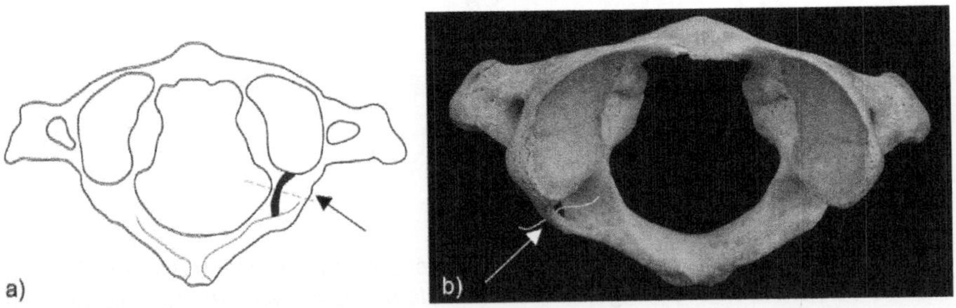

Figure 3.2. *Ponticulus posticus: (a) diagram, superior view; (b) dry bone, superior view. For a color version of this figure, see www.iste.co.uk/verna/asymptomatic.zip*

3.1.1.2. *Illustrations*

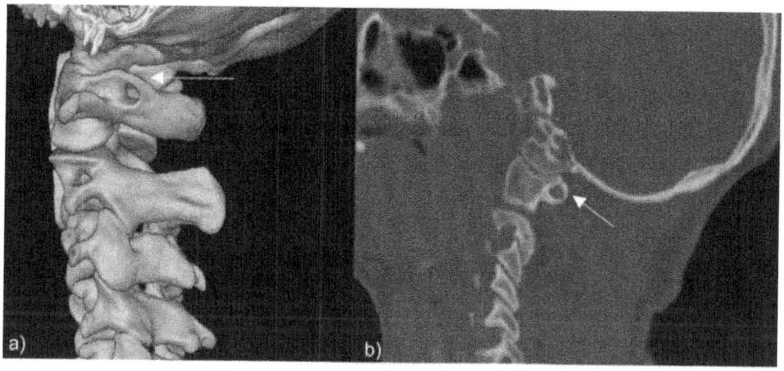

Figure 3.3. *Ponticulus posticus: (a) 3D reconstruction, left lateral view; (b) DICOM image, sagittal section. For a color version of this figure, see www.iste.co.uk/verna/asymptomatic.zip*

3.1.1.3. *Population frequencies*

Frequency	Population	References
8.7	Peruvian	(Perez et al. 2014)
9.1	Japanese	(Miki et al. 1979)
9.2	Chilean	(Schilling et al. 2010)
10.5	Korean	(Kim and Ratnayake 2015)
15.5	Korean	(Cho 2009)
16.7	Mediterranean	(Verna 2014)

Table 3.1. *Frequency of ponticulus posticus in medical imaging*

Frequency	Population	References
6.6	Indian (20th century)	(Hasan et al. 2001)
8.5	American (Eskimo, 16th–18th centuries)	(Saunders 1978)
8.5	American (8th–10th centuries)	(Saunders 1978)
8.6	Swiss (19th–20th centuries)	(Gemmerich 1999)
10.2	Greek (20th century)	(Paraskevas et al. 2005)
10.8	French (9th–12th centuries)	(Verna 2014)
12	French (1st–5th centuries)	(Murail 1996)
13.8	Polish (12th–13th centuries)	(Wysocki et al. 2003)
13.9	American (17th–18th centuries)	(Saunders 1978)
24	French (Neolithic)	(Thomas 2011)

Table 3.2. *Frequency of ponticulus posticus on dry bone*

3.1.1.4. *Association(s)*

A sex association was observed in two populations (Japanese and Korean populations). Men have a higher frequency of this variation than women. Concerning bilaterality, it depends on the populations. A more important unilaterality was observed in the Mediterranean and Korean populations, while the opposite was observed for the Swiss population (bilaterality being more observed).

3.1.2. *Lateral atlas bridge*

3.1.2.1. *Definition*

– Synonyms: None.

– Description: A complete or incomplete bony spicule located between the edge of the condylar facets and the transverse processes. The vertebral artery can pass through this bony bridge (Crubézy 1988).

3.1.2.2. *Illustrations*

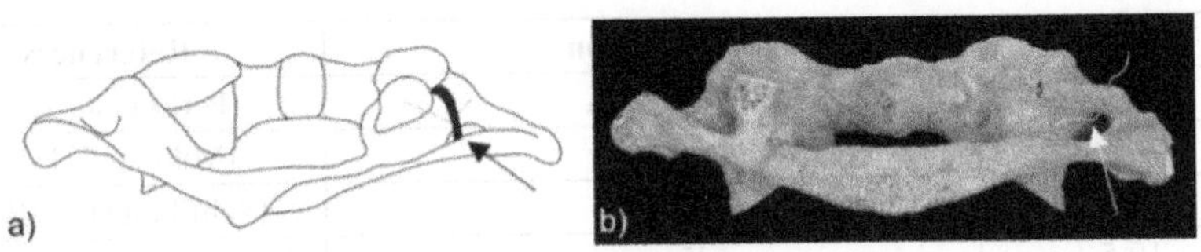

Figure 3.4. *Lateral atlas bridge: (a) diagram, posterior view; (b) dry bone, posterior view. For a color version of this figure, see www.iste.co.uk/verna/asymptomatic.zip*

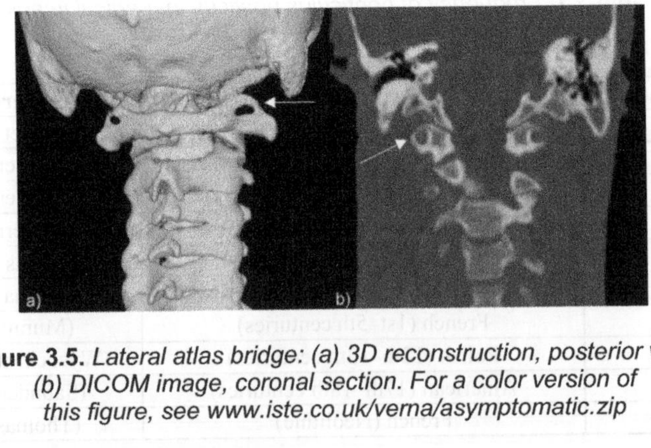

Figure 3.5. *Lateral atlas bridge: (a) 3D reconstruction, posterior view; (b) DICOM image, coronal section. For a color version of this figure, see www.iste.co.uk/verna/asymptomatic.zip*

3.1.2.3. *Population frequencies*

Frequency	Population	References
2.6	Mediterranean	(Verna 2014)

Table 3.3. *Frequency of lateral atlas bridges in medical imaging*

Frequency	Population	References
2	Indian (20th century)	(Hasan et al. 2001)
2.6	American (20th century)	(Karapetian 2017)
3	French (Neolithic)	(Thomas 2011)
4.2	Polish (12th–13th centuries)	(Wysocki et al. 2003)
8.3	American (17th–18th centuries)	(Saunders 1978)
9.9	American (Eskimo, 16th–18th centuries)	(Saunders 1978)
10.1	American (8th–10th centuries)	(Saunders 1978)
12.2	South African (19th–20th centuries)	(Mitchell 1998)

Table 3.4. *Frequency of lateral atlas bridges on dry bone*

3.1.2.4. *Association(s)*

The lateral bridge is mostly found unilaterally on the atlas.

3.1.3. *Retroarticular bridge*

3.1.3.1. *Definition*

– Synonyms: Retrotransverse canal and retro-transverse foramen.

– Description: It extends from the posterior border of the transverse process to the border of the posterior arch.

3.1.3.2. *Illustrations*

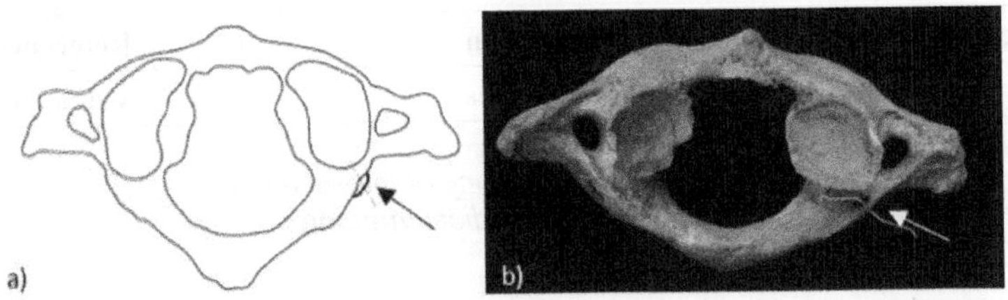

Figure 3.6. *Retroarticular bridge: (a) diagram, superior view; (b) dry bone, inferior view. For a color version of this figure, see www.iste.co.uk/verna/asymptomatic.zip*

3.1.3.3. *Population frequencies*

Frequency	Population	References
	No study	

Table 3.5. *Frequency of retroarticular bridges in medical imaging*

Frequency	Population	References
5	French (Neolithic)	(Thomas 2011)
8.3	American (17th–18th centuries)	(Saunders 1978)
9.7	French (9th–12th centuries)	(Verna 2014)
14.2	French	(Le Minor 1997)
14.7	American (8th–10th centuries)	(Saunders 1978)
14.7	Nepalese (20th century)	(Bilodi and Gupta 2005)
15.6	Swiss (19th–20th centuries)	(Gemmerich 1999)
18.4	American (Eskimo, 16th–18th centuries)	(Saunders 1978)

Table 3.6. *Frequency of retroarticular bridges on dry bone*

3.1.3.4. *Association(s)*

In the Swiss population, this variation is mostly found in men.

3.1.4. *Occipitalization*

3.1.4.1. *Definition*

– Synonym: Atlanto-occipital fusion.

– Description: It includes partial or complete fusion of the atlas with the foramen magnum.

This is due to craniocaudal border movement, failure to differentiate cranial and cervical parts (Barnes 2012).

3.1.4.2. *Illustrations*

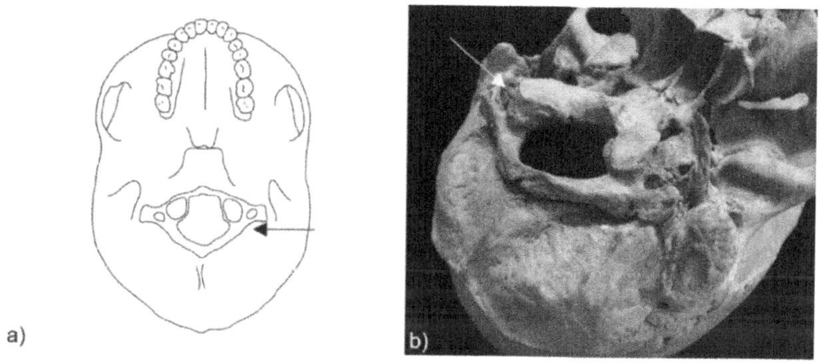

Figure 3.7. *Occipitalization of the atlas: (a) diagram, skull base, inferior view; (b) dry bone, skull base, inferior view. For a color version of this figure, see www.iste.co.uk/verna/asymptomatic.zip*

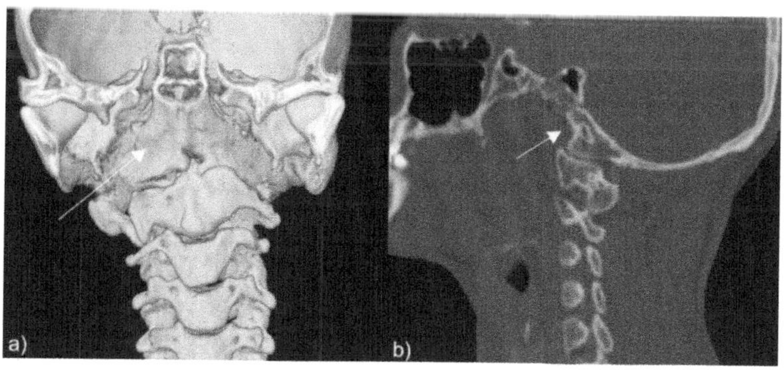

Figure 3.8. *Occipitalization of the atlas: (a) 3D reconstruction, anterior view; (b) DICOM image, sagittal section. For a color version of this figure, see www.iste.co.uk/verna/asymptomatic.zip*

3.1.4.3. *Population frequencies*

Frequency	Population	References
0.3	Mediterranean	(Verna 2014)
0.1	Korean	(Kim et al. 2015)

Table 3.7. *Frequency of occipitalization in medical imaging*

Frequency	Population	References
0	French (9th–12th centuries)	(Verna 2014)
Case report	Slovenian	(Dolensek et al. 2017)
0.6	Swiss (19th–20th centuries)	(Gemmerich 1999)
0.9	Saudi (20th century)	(Al-Motabagani and Surenda 2006)
1.6	Indian (20th century)	(Saini 2009)

Table 3.8. *Frequency of occipitalization on dry bone*

3.1.4.4. *Association(s)*

Bilaterality is more often observed for this variation.

3.1.5. *Anterior atlas arch deficiency*

3.1.5.1. *Definition*

– Synonyms: None.

– Description: It includes failure to fuse the anterior arch resulting in only a cleft, or failure of the ossification point to appear leading to the complete absence of the arch (Hosalkar et al. 2001).

3.1.5.2. *Illustrations*

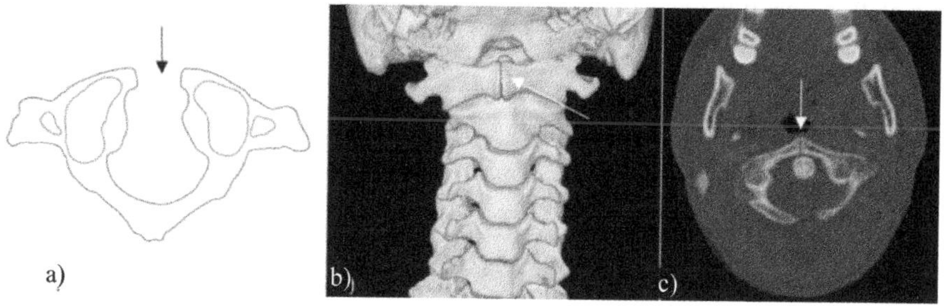

Figure 3.9. *Anterior atlas arch deficiency: (a) diagram, superior view; (b) 3D reconstruction, anterior view; (c) DICOM image, axial section. For a color version of this figure, see www.iste.co.uk/verna/asymptomatic.zip*

3.1.5.3. *Population frequencies*

Frequency	Population	References
0.1	Korean	(Kim et al. 2015)
0.2	Mediterranean	(Verna 2014)
0.5	American	(Senoglu et al. 2007)

Table 3.9. *Frequency of anterior atlas arch deficiency in medical imaging*

Frequency	Population	References
0	French (9th–12th centuries)	(Verna 2014)
1	Polish (12th–13th centuries)	(Wysocki et al. 2003)

Table 3.10. *Frequency of anterior atlas arch deficiency on dry bone*

3.1.5.4. *Association(s)*

No association was found for this AOV.

3.1.6. *Posterior atlas arch deficiency*

3.1.6.1. *Definition*

– Synonyms: None.

– Description: It occurs due to an ossification defect of the bone centers resulting in a cleft in one of the two arches or in the total absence of one or both arches.

3.1.6.2. *Illustrations*

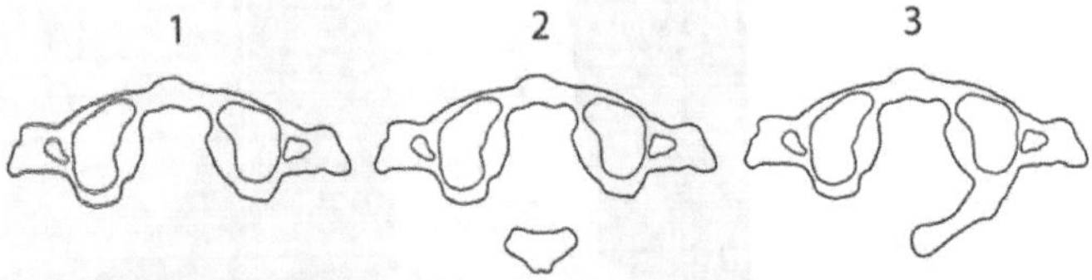

Figure 3.10. *Posterior atlas arch deficiency: diagram, three forms of the absence of the arch, superior view*

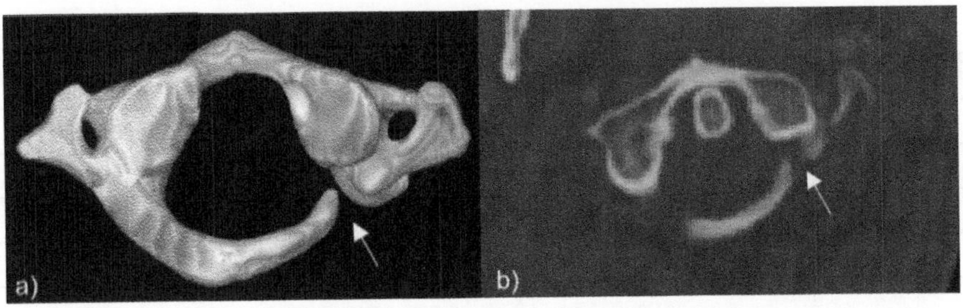

Figure 3.11. *Posterior atlas arch deficiency: (a) 3D reconstruction, inferior view; (b) DICOM image, axial section. For a color version of this figure, see www.iste.co.uk/verna/asymptomatic.zip*

3.1.6.3. *Population frequencies*

Frequency	Population	References
0.2	Mediterranean	(Verna 2014)
0.2	American	(Senoglu et al. 2007)
0.3	Korean	(Kim et al. 2015)
Case report	Turkish	(Ozdolap et al. 2007)

Table 3.11. *Frequency of posterior atlas arch deficiency in medical imaging*

Frequency	Population	References
0	French (9th–12th centuries)	(Verna 2014)

Table 3.12. *Frequency of posterior atlas arch deficiency on dry bone*

3.1.6.4. *Association(s)*

No association was found for this AOV.

3.1.7. *Double condylar facet*

3.1.7.1. *Definition*

– Synonym: Double facet form.

– Description: The upper facet of the atlas, articulating with that of the foramen magnum, may be double and separated by a ridge or a gutter.

3.1.7.2. *Illustrations*

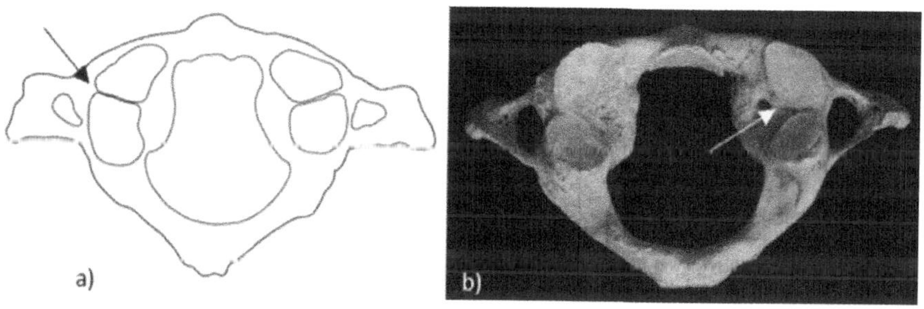

Figure 3.12. *Double condylar facet: (a) diagram, superior view; (b) dry bone, superior view. For a color version of this figure, see www.iste.co.uk/verna/asymptomatic.zip*

3.1.7.3. *Population frequencies*

Frequency	Population	References
No study		

Table 3.13. *Frequency of double condylar facets of the atlas in medical imaging*

Frequency	Population	References
5.8	French (4th–5th centuries)	(Castex 1994)
9	French (Neolithic)	(Thomas 2011)
10	French (4th–6th centuries)	(Castex 1994)
11	French (5th–7th centuries)	(Castex 1994)
12	French (5th–6th centuries)	(Castex 1994)
15	Polish (12th–13th centuries)	(Wysocki et al. 2003)
15.6	American (20th century)	(Karapetian 2017)
16.6	French (12th century)	(Castex 1994)
19.3	French (9th–12th centuries)	(Verna 2014)
31.8	Swiss (19th–20th centuries)	(Gemmerich 1999)

Table 3.14. *Frequency of double condylar facets of the atlas on dry bone*

3.1.7.4. *Association(s)*

Men are more affected than women (Gemmerich 1999; Thomas 2011) and unilaterality predominates (Castex 1994).

3.1.8. *Spina bifida occulta*

3.1.8.1. *Definition*

– Synonyms: None.

– Description: Due to a defect in the fusion of the bony lamina, the posterior arch of the atlas has a cleft.

3.1.8.2. *Illustrations*

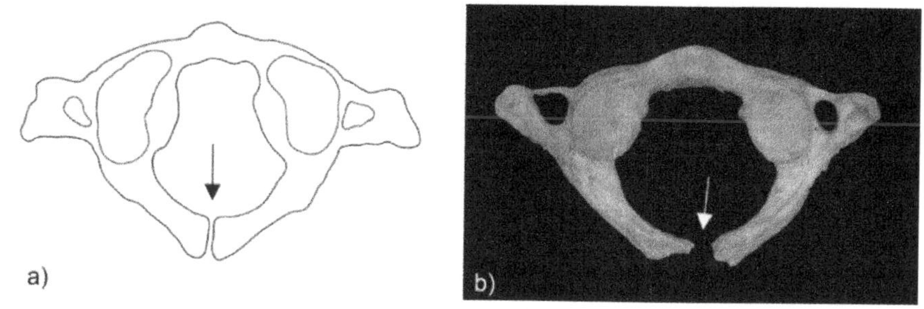

Figure 3.13. *Spina bifida occulta: (a) diagram, superior view; (b) dry bone, inferior view. For a color version of this figure, see www.iste.co.uk/verna/asymptomatic.zip*

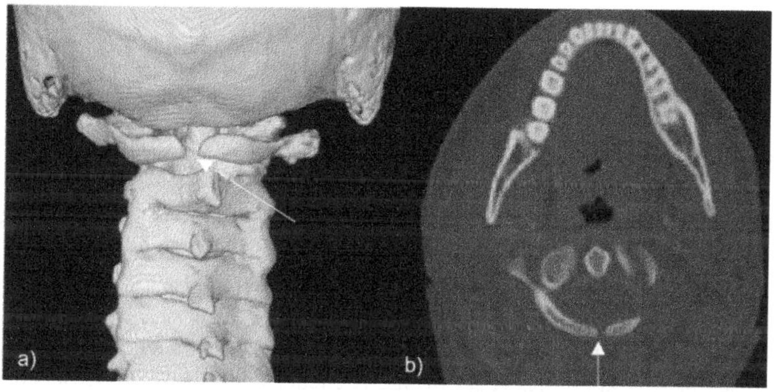

Figure 3.14. *Spina bifida occulta: (a) 3D reconstruction, posterior view; (b) DICOM image, axial section. For a color version of this figure, see www.iste.co.uk/verna/asymptomatic.zip*

3.1.8.3. *Population frequencies*

Frequency	Population	References
0.9	Korean	(Kim et al. 2015)
2.6	American	(Senoglu et al. 2007)
4.1	Mediterranean	(Verna 2014)

Table 3.15. *Frequency of spina bifida occulta in medical imaging*

Frequency	Population	References
0.3	American (Eskimo, 16th–18th centuries)	(Saunders 1978)
0.6	French (9th–12th centuries)	(Verna 2014)
1	Slovak (11th–12th centuries)	(Masnicova and Benus 2003)
1.1	French (5th–6th centuries)	(Castex 1994)
1.2	American (17th–18th centuries)	(Saunders 1978)
1.2	French (4th–17th centuries)	(Debono 2006)
2	Switzerland (19th–20th centuries)	(Gemmerich 1999)
2.4	American (8th–10th centuries)	(Saunders 1978)
3.2	Polish (12th–13th centuries)	(Wysocki et al. 2003)

Table 3.16. *Frequency of spina bifida occulta on dry bone*

3.1.8.4. *Association(s)*

Women have a higher frequency of this variation than men.

3.2. Second cervical vertebra (axis)

The second cervical vertebra is formed by six primary ossification centers, three of which form the tooth. The arch fuses around 3–4 years of age and fuses in the center around 4–6 years of age.

The two parts compose the tooth of the axis fuse at birth. The apical part of the tooth fuses last, at about 12 years of age (Scheuer and Black 2004; Scheuer et al. 2009). Failure of the tooth to fuse to the body can lead to an *os odontoideum*.

3.2.1. *Os odontoideum*

3.2.1.1. *Definition*

– Synonyms: None.

– Description: The odontoid process is separated from the vertebral body and is located in anatomical position.

3.2.1.2. *Illustrations*

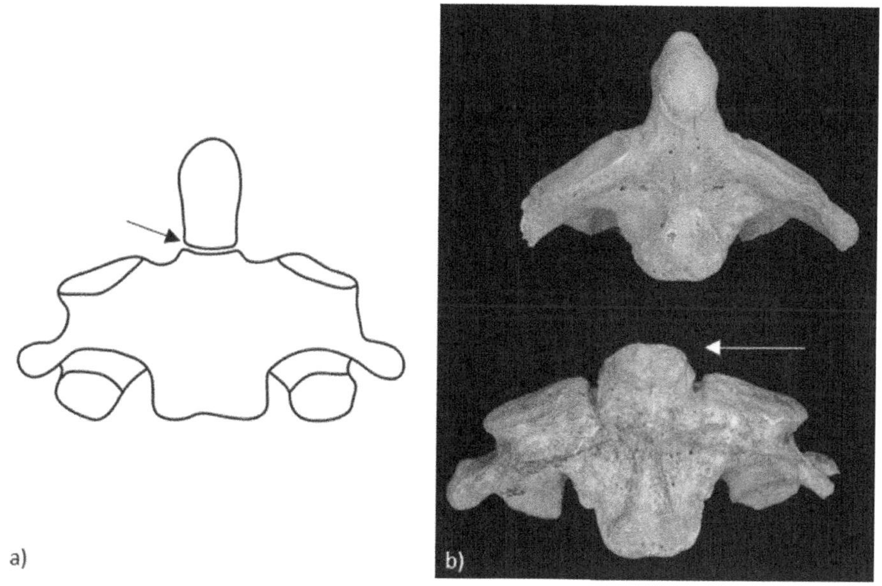

Figure 3.15. *Os odontoideum: (a) diagram, anterior view; (b) dry bone, anterior view (normal axis on top, axis without os odontoideum on the bottom). For a color version of this figure, see www.iste.co.uk/verna/asymptomatic.zip*

3.2.1.3. *Population frequencies*

Frequency	Population	References
0	Mediterranean	(Verna 2014)
Case report	Turkish	(Altun and Yuksel 2017)

Table 3.17. *Frequency of os odontoideum in medical imaging*

Frequency	Population	References
0.6	French (9th–12th centuries)	(Verna 2014)
1.3	Portuguese (13th–14th centuries)	(Curate 2008)

Table 3.18. *Frequency of os odontoideum on dry bone*

3.2.1.4. *Association(s)*

No association was found for this AOV.

3.3. Cervical vertebra

The vertebrae are formed from three primary ossification centers: one for the body and two for the vertebral arches.

The two arches fuse at the level of the spinous process around 1–2 years of age; a lack of fusion can lead to *spina bifida occulta*. The epiphyses appear at puberty, and the vertebrae are mature by 20 years of age (Scheuer and Black 2004; Scheuer et al. 2009).

The cervical vertebrae show three AOVs, all of which can be observed from about 10 years of age, when they are fused (body and arches).

3.3.1. *Open transverse foramen*

3.3.1.1. *Definition*

– Synonyms: None.

– Description: It occurs due to a fusion defect of the transverse foramen resulting in an incomplete foramen.

3.3.1.2. *Illustrations*

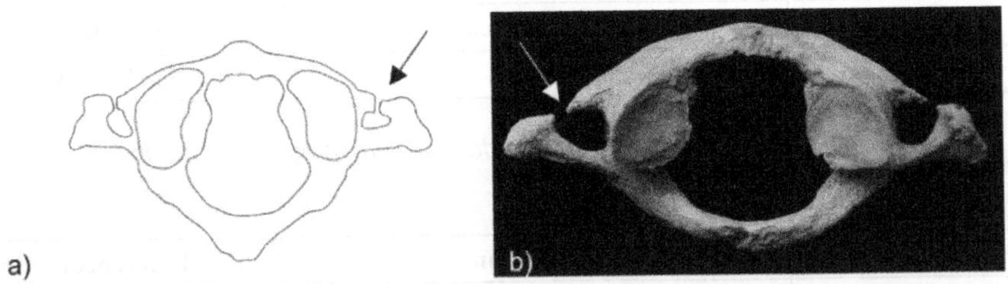

Figure 3.16. *Open transverse foramen: (a) diagram of first cervical vertebra, superior view; (b) dry bone, first cervical vertebra, inferior view. For a color version of this figure, see www.iste.co.uk/verna/asymptomatic.zip*

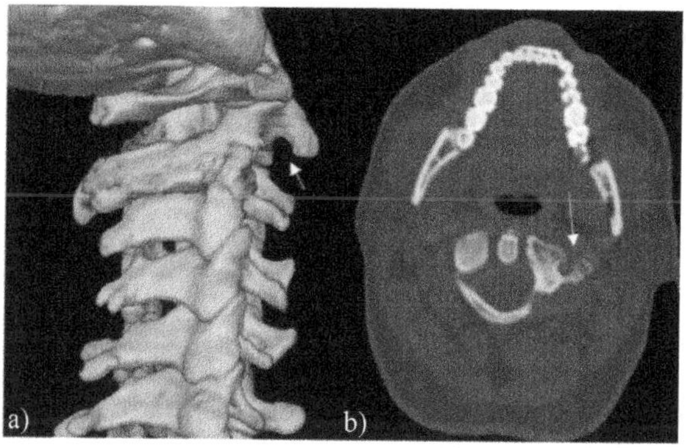

Figure 3.17. *Open transverse foramen: (a) 3D reconstruction, second cervical vertebra, left lateral view; (b) DICOM image, first cervical vertebra, axial section. For a color version of this figure, see www.iste.co.uk/verna/asymptomatic.zip*

3.3.1.3. *Population frequencies*

Incomplete transverse foramen can be seen on all cervical vertebrae, but the first cervical vertebra is the most affected. In general, the frequency is lower on dry bone than on imaging and this may be due to preservation problems. This is a relatively fragile area, and it is important to differentiate between lack of fusion of the arch and breakage.

Frequency	Population	References
8.5	Mediterranean	(Verna 2014)

Table 3.19. *Frequency of open transverse foramen in medical imaging*

Frequency	Population	References
0	French (19th–20th centuries)	(Verna 2014)
1.5	French (9th–12th centuries)	(Verna 2014)
5	Polish (12th–13th centuries)	(Wysocki et al. 2003)
5	French (Neolithic)	(Thomas 2011)
10	Indian (20th century)	(Chauhan and Khanna 2012)

Table 3.20. *Frequency of open transverse foramen on dry bone*

3.3.1.4. *Association(s)*

Open transverse foramen is more often found bilaterally than unilaterally.

3.3.2. *Double transverse foramen*

3.3.2.1. *Definition*

– Synonyms: None.

– Description: It includes transverse foramen separated into two by a thin bony lamina giving two distinct foramina.

3.3.2.2. *Illustrations*

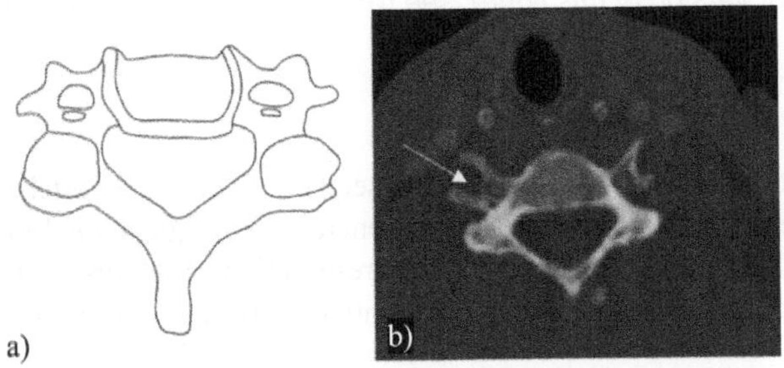

Figure 3.18. *Double transverse foramen: (a) diagram, superior view; (b) DICOM image, axial section*

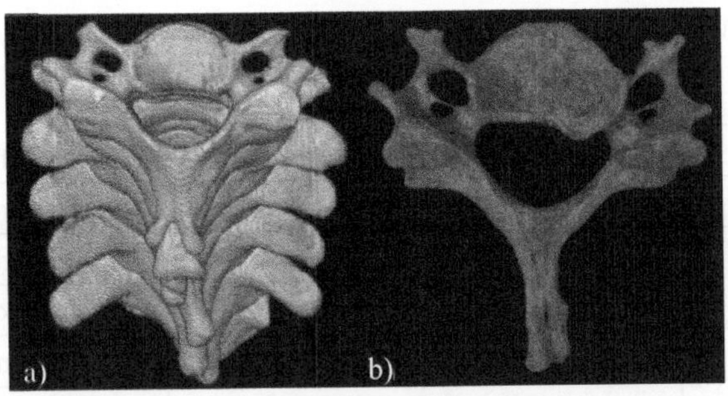

Figure 3.19. *Double transverse foramen: (a) 3D reconstruction, superior view; (b) dry bone, superior view. For a color version of this figure, see www.iste.co.uk/verna/asymptomatic.zip*

3.3.2.3. *Population frequencies*

The second and seventh cervical vertebrae show the lowest frequencies (between 2% and 5%), and the sixth one has the highest frequency (50%). This variation is never observed on the atlas.

Vertebrae	Frequency	Population	References
C2	0	Mediterranean	(Verna 2014)
C3	0.3		
C4	4.9		
C5	27.3		
C6	53.5		
C7	5		

Table 3.21. *Frequency of double transverse foramen in medical imaging*

Vertebrae	Frequency	Population	References
C2	0.3	Swiss (19th–20th centuries)	(Gemmerich 1999)
C3	0.7	American (20th century)	(Karapetian 2017)
	2.6	French (19th–20th centuries)	(Verna 2014)
	2.8	Polish (12th–13th centuries)	(Wysocki et al. 2003)
C4	5.9	French (19th–20th centuries)	(Verna 2014)
	11	American (20th century)	(Karapetian 2017)
	13.5	Polish (12th–13th centuries)	(Wysocki et al. 2003)
C5	24.4	Polish (12th–13th centuries)	(Wysocki et al. 2003)
	26	American (20th century)	(Karapetian 2017)
	33.3	French (19th–20th centuries)	(Verna 2014)
C6	45.6	Polish (12th–13th centuries)	(Wysocki et al. 2003)
	46	American (20th century)	(Karapetian 2017)
	52.4	French (19th–20th centuries)	(Verna 2014)
C7	0	French (19th–20th centuries)	(Verna 2014)
	14	American (20th century)	(Karapetian 2017)

Table 3.22. *Frequency of double transverse foramen on dry bone*

3.3.2.4. *Association(s)*

A preference for the left side and unilaterality is found as well as a preference in women for the sixth and seventh cervical vertebrae. The frequency increases somewhat with age, and Castex (1994) shows a preference for bilaterality of this AOV.

3.3.3. *Bifid spinous process*

3.3.3.1. *Definition*

– Synonyms: None.

– Description: It includes a spiny process with two ends.

3.3.3.2. *Illustrations*

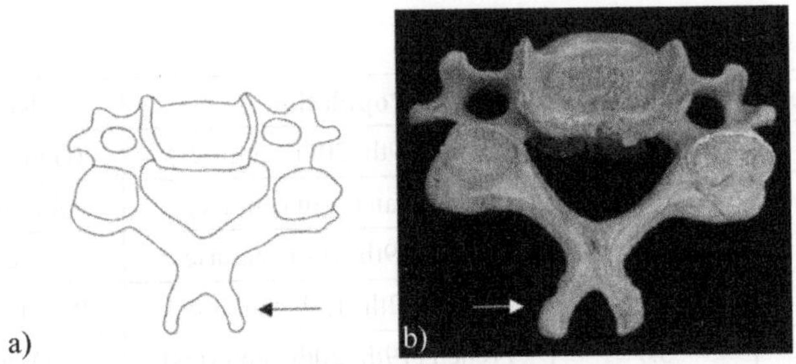

Figure 3.20. *Bifid spinous process: (a) diagram, superior view; (b) dry bone, superior view. For a color version of this figure, see www.iste.co.uk/verna/asymptomatic.zip*

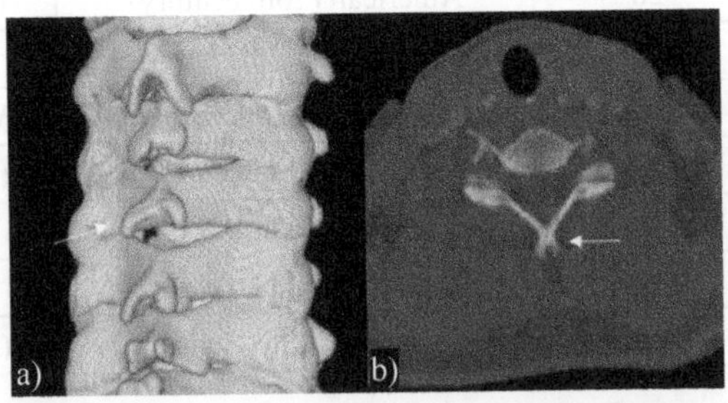

Figure 3.21. *Bifid spinous process: (a) 3D reconstruction, posterior view; (b) DICOM image, axial section. For a color version of this figure, see www.iste.co.uk/verna/asymptomatic.zip*

3.3.3.3. *Population frequencies*

It has the highest frequency at the level of the second cervical vertebra and the lowest for the seventh, which is closer in its conformation to a thoracic vertebra.

Vertebrae	Frequency	Population	References
C2	95.4	Mediterranean	(Verna 2014)
C3	64.2		
C4	84.1		
C5	89.9		
C6	54.6		
C7	1.4		

Table 3.23. *Frequency of bifid spinous process in medical imaging*

Vertebrae	Frequency	Population	References
C2	91	American (20th century)	(Duray et al. 1999)
C3	36	American (20th century)	(Duray et al. 1999)
	73.4	French (19th–20th centuries)	(Verna 2014)
C4	51	American (20th century)	(Duray et al. 1999)
	85.4	French (19th–20th centuries)	(Verna 2014)
C5	66.2	American (20th century)	(Duray et al. 1999)
	93.6	French (19th–20th centuries)	(Verna 2014)
C6	42.7	American (20th century)	(Duray et al. 1999)
	50.6	French (19th–20th centuries)	(Verna 2014)
C7	0	French (19th–20th centuries)	(Verna 2014)
	0.8	American (20th century)	(Duray et al. 1999)

Table 3.24. *Frequency of bifid spinous process on dry bone*

3.3.3.4. *Association(s)*

This variation is more present in men than in women.

3.4. Spine

3.4.1. *Spina bifida occulta*

3.4.1.1. *Definition*

– Synonyms: None.

– Description: Due to a defect in the fusion of the bone plate, the posterior arch of the vertebra is separated in two.

3.4.1.2. *Illustrations*

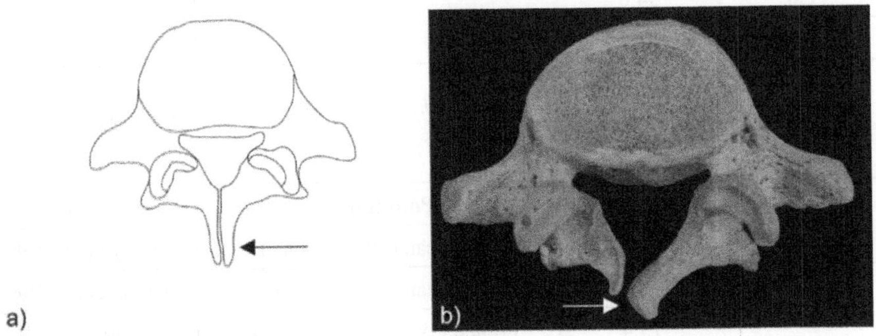

Figure 3.22. *Spina bifida occulta: (a) diagram, lumbar vertebra, superior view; (b) dry bone, lumbar vertebra, superior view. For a color version of this figure, see www.iste.co.uk/verna/asymptomatic.zip*

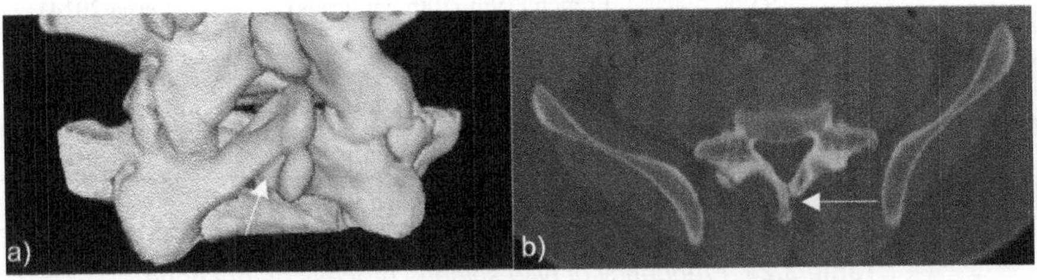

Figure 3.23. *Spina bifida occulta: (a) 3D reconstruction, lumbar vertebra, posterior view; (b) DICOM image, lumbar vertebra, axial section. For a color version of this figure, see www.iste.co.uk/verna/asymptomatic.zip*

3.4.1.3. *Population frequencies*

Spina bifida occulta can affect the entire vertebral column. On the other hand, problems with the preservation of the bones may prevent an optimal scoring and decrease the frequency. The most affected vertebra is the atlas, and it is also the most studied vertebra, both in imaging and on dry bone.

Frequency	Population	References
0.7	English	(Saluja 1988)
6.6	Mediterranean	(Verna 2014)

Table 3.25. *Frequency of spina bifida occulta in all vertebrae combined in medical imaging*

Frequency	Population	References
2	American	(Saunders 1978)
3.5	French (19th–20th centuries)	(Verna 2014)
3.8	French (9th–12th centuries)	(Verna 2014)

Table 3.26. *Frequency of spina bifida occulta in all vertebrae combined on dry bone*

3.4.1.4. *Association(s)*

No association was found for this AOV.

3.4.2. *Butterfly vertebra*

3.4.2.1. *Definition*

– Synonyms: None.

– Description: It occurs due to a defect in the notochord during vertebral segment development that can lead to a bifid center (sagittal cleft in the center) resulting in a so-called "butterfly" vertebra (Barnes 2012).

3.4.2.2. *Illustrations*

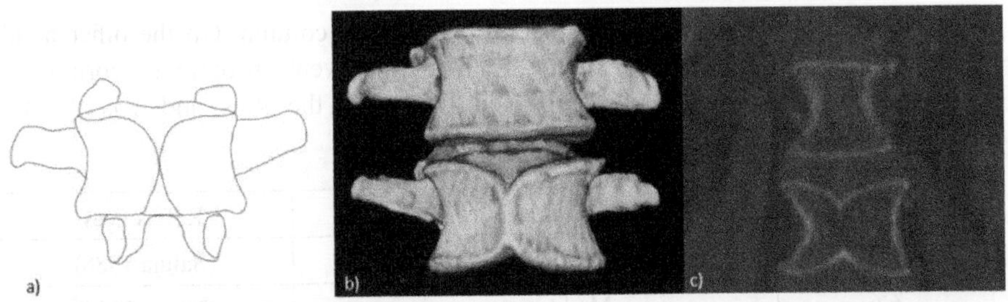

Figure 3.24. *Butterfly vertebra: (a) diagram, lumbar vertebra, anterior view; (b) 3D reconstruction, lumbar vertebra, anterior view; (c) DICOM image, coronal section. For a color version of this figure, see www.iste.co.uk/verna/asymptomatic.zip*

3.4.2.3. *Population frequencies*

In imaging, the majority of cases are found in the form of case reports, so it is difficult to obtain a frequency of this AOV and to know in what proportion the authors observe a so-called "mild" form affecting mainly the first sacral vertebra or the butterfly wing.

Frequency	Population	References
0.1	Mediterranean	(Verna 2014)
Case report	Turkish	(Sonel et al. 2001)
Case report	Tunisian	(Slouma et al. 2016)

Table 3.27. *Frequency of the butterfly vertebrae in all vertebrae combined in medical imaging*

A significant frequency is observed for the Inuit population. An explanation may lie in the genetic origin of this variation.

A population with less genetic mixing may find this type of variation in a higher proportion.

Frequency	Population	References
0	French (19th–20th centuries)	(Verna 2014)
0	French (9th–12th centuries)	(Verna 2014)
1.6	Italian (20th century)	(Brasili et al. 2002)
12	American (Inuit, 12th–19th centuries)	(Merbs 2004)

Table 3.28. *Frequency of butterfly vertebrae in all vertebrae combined on dry bone*

3.4.2.4. *Association(s)*

No association was found for this AOV.

3.4.3. *Spondylolysis*

3.4.3.1. *Definition*

– Synonyms: None.

– Description: It includes the vertebral separation at the level of the isthmus separating the upper and lower articular processes (*pars interarticularis*), and separation of the neural arch from the rest of the vertebra.

This is due to a fatigue fracture: repeated hyper-extension, hyper-flexion or rotation of the spine.

3.4.3.2. *Illustrations*

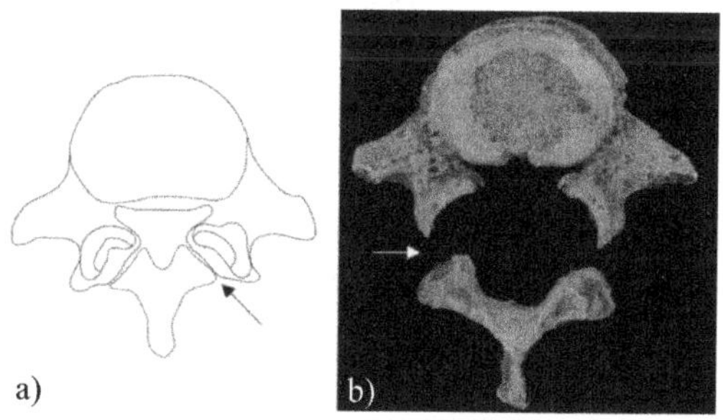

Figure 3.25. *Spondylolysis: (a) diagram, lumbar vertebra, superior view; (b) dry bone, lumbar vertebra, superior view. For a color version of this figure, see www.iste.co.uk/verna/asymptomatic.zip*

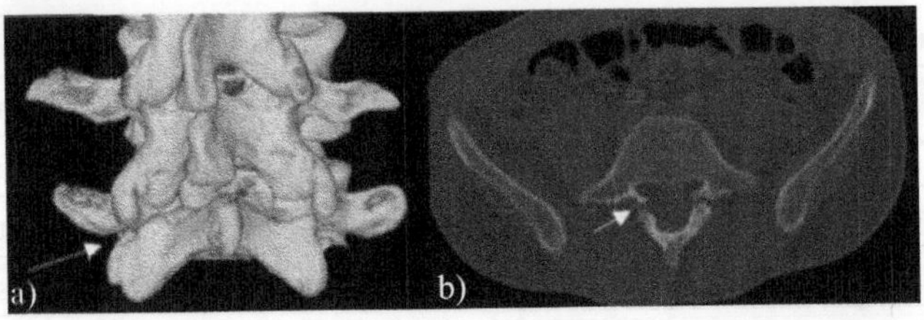

Figure 3.26. *Spondylolysis: (a) 3D reconstruction, lumbar vertebra, posterior view; (b) DICOM image, axial section. For a color version of this figure, see www.iste.co.uk/verna/asymptomatic.zip*

3.4.3.3. *Population frequencies*

Frequency	Population	References
5.2	Mediterranean	(Verna 2014)

Table 3.29. *Frequency of spondylolysis in all vertebrae combined in medical imaging*

Frequency	Population	References
3.2	French (4th–5th centuries)	(Castex 1994)
4	Slovak (9th century)	(Masnicova and Benus 2003)
4.8	French (9th–12th centuries)	(Verna 2014)
4.9	French (4th–17th centuries)	(Debono 2006)
5.9	French (19th–20th centuries)	(Verna 2014)
6.6	French (8th–12th centuries)	(Crubézy 1988)
7	Slovak (9th–12th centuries)	(Masnicova and Benus 2003)
11.5	French (4th–6th centuries)	(Castex 1994)
13.8	American (8th–10th centuries)	(Saunders 1978)
22.3	American (17th–18th centuries)	(Saunders 1978)
27.5	American (Eskimo, 16th–18th centuries)	(Saunders 1978)
33	American (Eskimo)	(Stewart 1953)

Table 3.30. *Frequency of spondylolysis in all vertebrae combined on dry bone*

Spondylolysis may be present in the entire vertebral column, but it is found preferentially in the fifth lumbar vertebra. A few cases have been observed in the fourth lumbar vertebra or in the lumbosacral transitional vertebra.

3.4.3.4. *Association(s)*

Spondylolysis is predominantly present in men and bilaterally at the isthmus.

3.4.4. *Transitional vertebra*

3.4.4.1. *Definition*

– Synonyms: None.

– Description: An extra vertebra at the thoracolumbar junction or at the lumbosacral junction may be present.

It is difficult to make a clear distinction between a 13th thoracic vertebra or a sixth lumbar vertebra; it seems more prudent to call it a transitional vertebra to avoid mistakes.

3.4.4.2. *Illustrations*

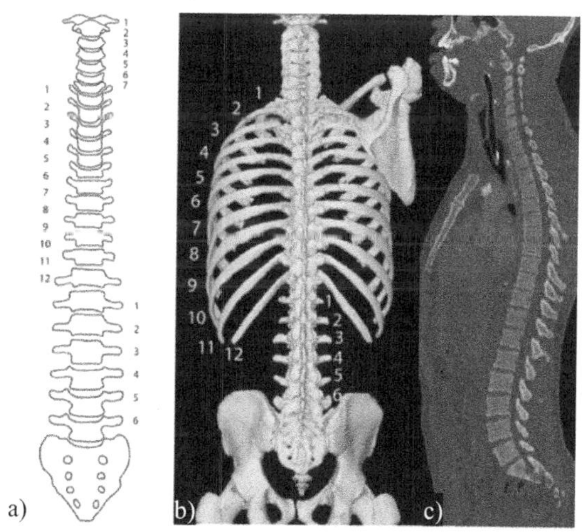

Figure 3.27. *Transitional vertebra: (a) diagram, anterior view; (b) 3D reconstruction, posterior view; (c) DICOM image, sagittal section. For a color version of this figure, see www.iste.co.uk/verna/asymptomatic.zip*

3.4.4.3. *Population frequencies*

Frequency	Population	References
6.6	American	(Apazidis et al. 2011)
10.6	Mediterranean	(Verna 2014)

Table 3.31. *Frequency of transitional vertebrae in medical imaging*

Frequency	Population	References
4.2	American (20th century)	(Tague 2011)
5.2	French (4th–5th centuries)	(Castex 1994)
5.6	American (20th century)	(Tague 2017)
10.8	French (9th–12th centuries)	(Verna 2014)
12.9	French (5th–7th centuries)	(Castex 1994)
24.4	French (19th–20th centuries)	(Verna 2014)

Table 3.32. *Frequency of transitional vertebrae on dry bone*

3.4.4.4. *Association(s)*

No association was found for this AOV.

3.4.5. *Sacralization*

3.4.5.1. *Definition*

– Synonyms: None.

– Description: It includes fusion of the fifth lumbar vertebra with the first sacral vertebra in a symmetrical or asymmetrical way.

This is due to cranial movement (Barnes 2012).

3.4.5.2. *Illustrations*

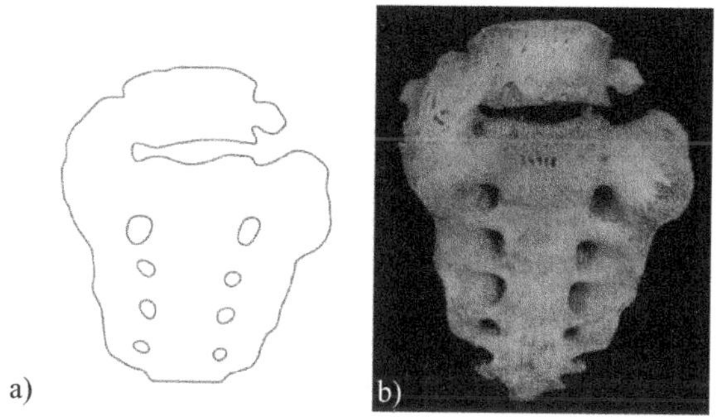

Figure 3.28. *Sacralization: (a) diagram, anterior view; (b) dry bone anterior view. For a color version of this figure, see www.iste.co.uk/verna/asymptomatic.zip*

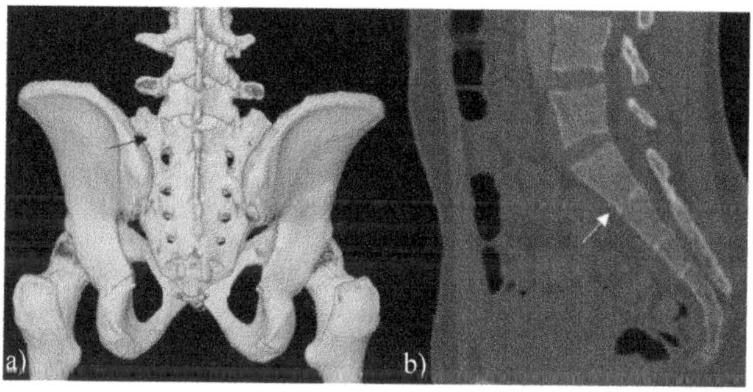

Figure 3.29. *Sacralization: (a) 3D reconstruction, posterior view; (b) DICOM image, sagittal section. For a color version of this figure, see www.iste.co.uk/verna/asymptomatic.zip*

3.4.5.3. *Population frequencies*

Frequency	**Population**	**References**
2.7	Japanese	(Nakajima et al. 2014)
6.7	Mediterranean	(Verna 2014)
8	American	(Apazidis et al. 2011)
13.9	Korean	(Lee 2007)

Table 3.33. *Frequency of sacralization in medical imaging*

Frequency	Population	References
0.7	Egyptian (4th–6th dynasties)	(El-Din and El Banna 2006)
6.1	American	(Willis 1929)
7	Slovak (9th century)	(Masnicova and Benus 2003)
8	Slovak (11th–12th centuries)	(Masnicova and Benus 2003)
8.8	French (9th–12th centuries)	(Verna 2014)
10.2	American (20th century)	(Tague 2011)
17.4	French (19th–20th centuries)	(Verna 2014)

Table 3.34. *Frequency of sacralization on dry bone*

3.4.5.4. *Association(s)*

Some sacralizations may be due to age with fusion of the wings (rather pathological cases that may have been wrongly sized and thus increasing the frequency). It is therefore necessary to be vigilant when rating this AOV.

The study by Masnicova and Benus (2003) highlights a higher frequency of sacralization in men.

4

The Pelvic Girdle

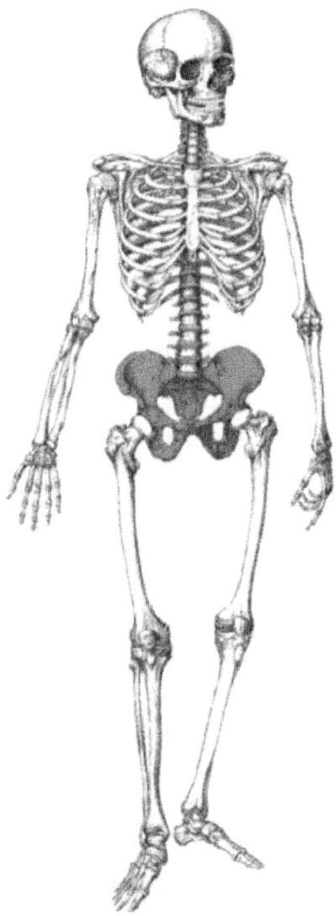

Figure 4.1. *Human skeleton. The pelvic girdle is presented in red. For a color version of this figure, see www.iste.co.uk/verna/asymptomatic.zip*

The pelvic girdle, composed of the sacrum and the coxal bones, has seven AOVs, which will be described in this chapter. Most of the studies performed on the pelvic girdle are based on dry bones, and this area has been less studied than the rest of the skeleton.

4.1. Sacrum

The sacrum is made up of five sacral vertebrae, each composed of three primary ossification centers (one for the body, two for the arch), except for the first two sacral vertebrae, which are formed by two additional centers for the lateral elements. Each vertebra forms only one entity between 2 and 6 years of age. Then, they fuse together starting at age 12, beginning at the distal end, moving up from S5, and ending with S1–S2 by age 25 (Scheuer and Black 2004; Scheuer et al. 2009). Fusion of the posterior arches occurs between 7 and 15 years of age; a lack of fusion leads to *spina bifida occulta*. Four AOVs are observable for the sacrum. In order to be sure that an anatomical variation is present, for lumbarization and *spina bifida occulta*, it is necessary to wait for complete maturation of the bone.

4.1.1. *Lumbarization*

4.1.1.1. *Definition*

– Synonyms: None.

– Description: The first sacral vertebra becomes a transitional vertebra, and thus it is individualized from the sacrum.

This is due to caudal displacement at the lumbosacral junction (Barnes 2012).

4.1.1.2. *Illustrations*

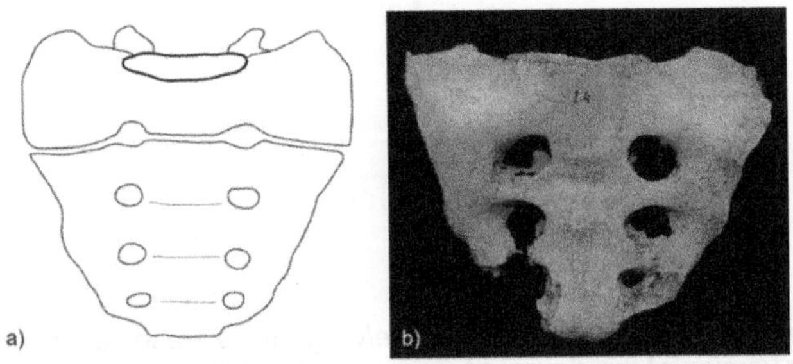

Figure 4.2. *Lumbarization: (a) diagram, anterior view; (b) dry bone, anterior view. For a color version of this figure, see www.iste.co.uk/verna/asymptomatic.zip*

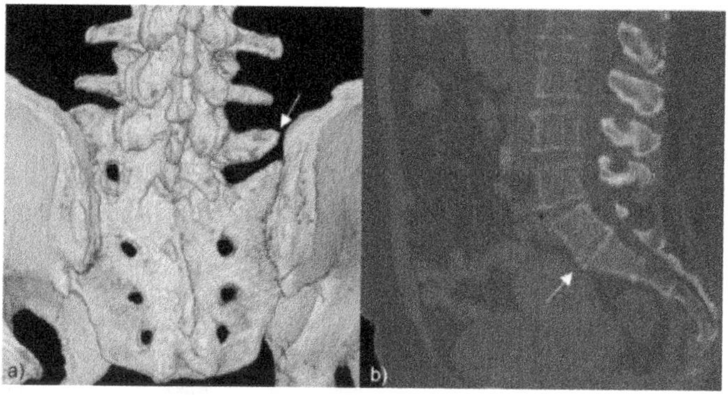

Figure 4.3. *Lumbarization: (a) 3D reconstruction, posterior view; (b) DICOM image, sagittal section. For a color version of this figure, see www.iste.co.uk/verna/asymptomatic.zip*

4.1.1.3. *Population frequencies*

Frequency	Population	References
2.2	Mediterranean	(Verna 2014)
6.2	Japanese	(Nakajima et al. 2014)
9.9	Korean	(Lee et al. 2007)

Table 4.1. *Frequency of lumbarization in medical imaging*

Frequency	Population	References
1.2	French (19th–20th centuries)	(Verna 2014)
1.5	Egyptian (4th–6th dynasties)	(El-Din and El Banna 2006)
1.8	French (9th–12th centuries)	(Verna 2014)
2	Slovak (11th–12th centuries)	(Masnicova and Benus 2003)
3.9	Indian (20th century)	(Mahato 2010)

Table 4.2. *Frequency of lumbarization on dry bone*

4.1.1.4. *Association(s)*

No association was found for this AOV.

4.1.2. *Spina bifida occulta*

4.1.2.1. *Definition*

– Synonyms: None.

– Description: Due to a defect in the fusion of the bony lamina, the posterior arch of the sacrum is separated in two. It may be complete or incomplete.

4.1.2.2. *Illustrations*

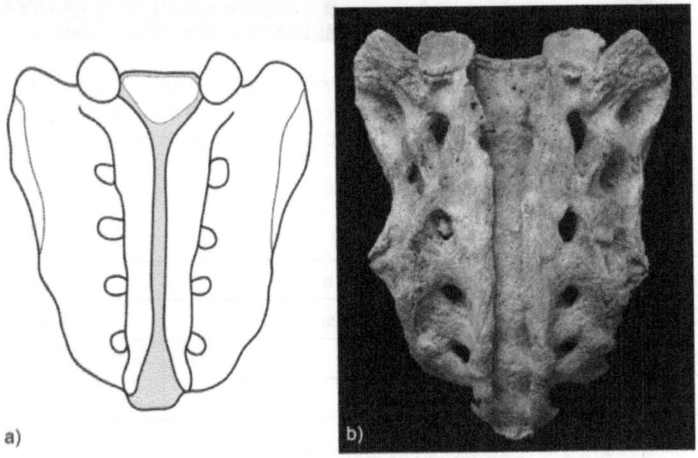

Figure 4.4. *Complete spina bifida occulta: (a) diagram, posterior view; (b) dry bone, posterior view. For a color version of this figure, see www.iste.co.uk/verna/asymptomatic.zip*

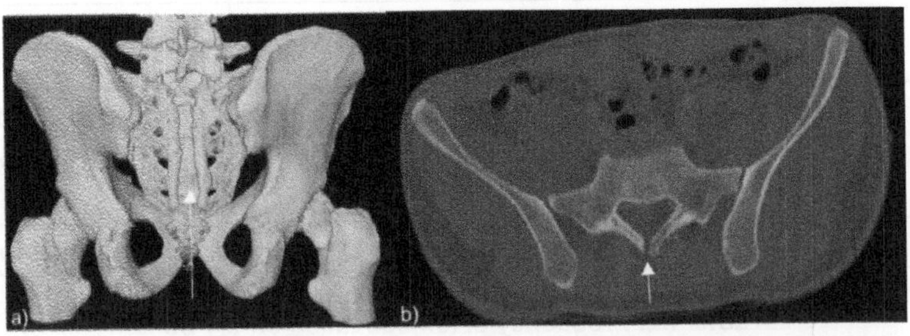

Figure 4.5. *Complete spina bifida occulta: (a) 3D reconstruction, posterior view; (b) DICOM image, axial section. For a color version of this figure, see www.iste.co.uk/verna/asymptomatic.zip*

4.1.2.3. *Population frequencies*

Spina bifida occulta affects the entire sacrum, with a high frequency at the level of the first sacral vertebra and a lower frequency when the entire sacrum is involved. Its overall frequency is 35%.

Frequency	Population	References
15	English	(Saluja 1988)
22	English	(Boone et al. 1985)
35	Mediterranean	(Verna 2014)
66	Australian	(Albrecht et al. 2007)

Table 4.3. *Frequency of spina bifida occulta of the sacrum in all vertebrae combined in medical imaging*

Frequency	Population	References
3.2	French (5th–7th centuries)	(Castex 1994)
3.3	Egyptian (4th–6th dynasties)	(El-Din and El Banna 2006)
13	French (Neolithic)	(Thomas 2011)
15.1	English (18th–19th centuries)	(Saluja 1988)
20.8	French (4th–17th centuries)	(Debono 2006)
23	Slovak (9th century)	(Masnicova and Benus 2003)
23.8	Slovak (11th–12th centuries)	(Masnicova and Benus 2003)
28.1	Chinese (20th century)	(Wu et al. 2009)
35.4	French (9th–12th centuries)	(Verna 2014)

Table 4.4. *Frequency of spina bifida occulta of the sacrum in all vertebrae combined on dry bone*

4.1.2.4. *Association(s)*

It has a higher frequency in men and appears to be more common in those under 40.

4.1.3. *Butterfly vertebra*

4.1.3.1. *Definition*

– Synonyms: None.

– Description: It includes a slight depression in the center of the vertebral body forming a gutter.

4.1.3.2. *Illustrations*

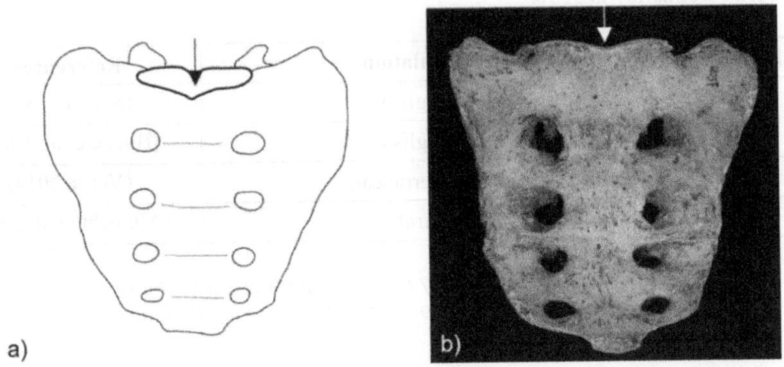

Figure 4.6. *Butterfly vertebra: (a) diagram, anterior view; (b) dry bone, anterior view. For a color version of this figure, see www.iste.co.uk/verna/asymptomatic.zip*

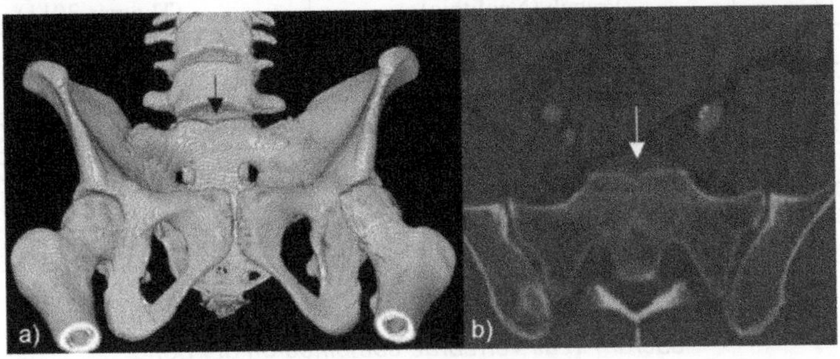

Figure 4.7. *Butterfly vertebra: (a) 3D reconstruction, infero-anterior view; (b) DICOM image, axial section. For a color version of this figure, see www.iste.co.uk/verna/asymptomatic.zip*

4.1.3.3. *Population frequencies*

Frequency	Population	References
2.6	Mediterranean	(Verna 2014)

Table 4.5. *Frequency of butterfly vertebra of the first sacral vertebra in medical imaging*

Frequency	Population	References
6.1	French (19th–20th centuries)	(Verna 2014)
6.1	French (9th–12th centuries)	(Verna 2014)

Table 4.6. *Frequency of butterfly vertebra of the first sacral vertebra on dry bone*

4.1.3.4. *Association(s)*

No association was found for this AOV.

4.1.4. *Accessory hip facet*

4.1.4.1. *Definition*

– Synonym: Accessory sacroiliac facet.

– Description: It is ovoid in shape and located at the level of the first or the second sacral foramen.

The carrying of heavy loads on the back may be responsible for its appearance, inducing flexion of the spine (Capasso et al. 1999).

4.1.4.2. *Illustrations*

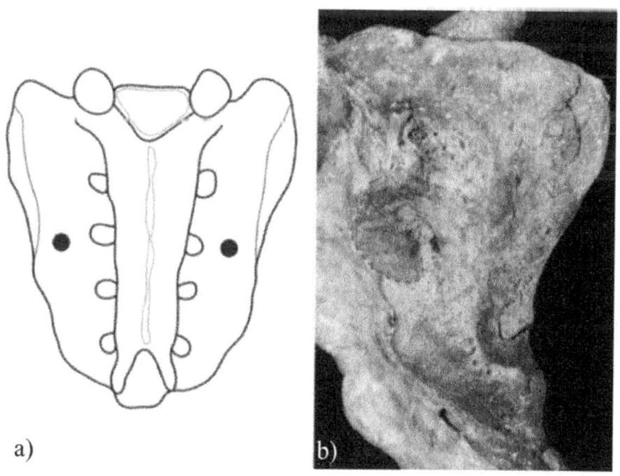

Figure 4.8. *Accessory hip facet: (a) diagram, posterior view; (b) dry bone, right lateral view. For a color version of this figure, see www.iste.co.uk/verna/asymptomatic.zip*

4.1.4.3. *Population frequencies*

Frequency	Population	References
	No study	

Table 4.7. *Frequency of accessory hip facet in medical imaging*

Frequency	Population	References
14.9	American (Eskimo, 16th–18th centuries)	(Saunders 1978)
16.1	American (20th century)	(Trotter 1937)
16.2	American (8th–10th centuries)	(Saunders 1978)
21.2	French (9th–12th centuries)	(Verna 2014)
24	American (17th–18th centuries)	(Saunders 1978)

Table 4.8. *Frequency of accessory hip facet on dry bone*

4.1.4.4. *Association(s)*

It is mostly found in women (Trotter 1937), which is not the case in other studies. On the other hand, a preference for the right side in the case of unilaterality is observed; otherwise, on the whole, bilaterality is more significant.

4.2. Coxal bone

The coxal bone forms from three primary ossification centers representing the ilium, pubis and ischium. Five secondary ossification centers make up the epiphyses, which appear between 12 and 20 years of age and fuse between 18 and 25 years of age.

The three bones comprising the coxal bone fuse at the Y-shaped cartilage from age 11 to 17 (at puberty, so a lag is observed between males and females) (Scheuer and Black 2004; Scheuer et al. 2009).

Regarding the coxal bone, three AOVs can be observed. The acetabular mark should be sized after the fusion between the three bones composing the coxal bone, that is, at the end of puberty.

4.2.1. *Accessory hip facet*

4.2.1.1. *Definition*

– Synonym: Accessory sacroiliac facet.

– Description: It is ovoid in shape and located at the level of the iliac tuberosity of the hip bone.

Carrying heavy loads on the back may be responsible for occurrence, inducing flexion of the spine (Capasso et al. 1999).

4.2.1.2. *Illustrations*

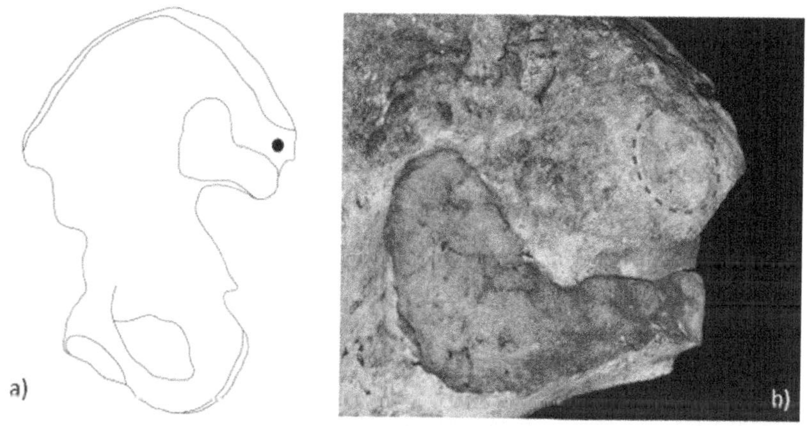

Figure 4.9. *Accessory hip facet: (a) diagram, medial view; (b) dry bone, medial view. For a color version of this figure, see www.iste.co.uk/verna/asymptomatic.zip*

4.2.1.3. *Population frequencies*

Frequency	Population	References
	No study	

Table 4.9. *Frequency of accessory hip facet on the coxal bone in medical imaging*

Frequency	Population	References
16.1	American (20th century)	(Trotter 1937)
18.2	French (9th–12th centuries)	(Verna 2014)
29.7	French (19th–20th centuries)	(Verna 2014)
34.6	American (Eskimo, 16th–18th centuries)	(Saunders 1978)
36.4	American (8th–10th centuries)	(Saunders 1978)
44.1	American (17th–18th centuries)	(Saunders 1978)

Table 4.10. *Frequency of accessory hip facet on the coxal bone on dry bone*

4.2.1.4. *Association(s)*

The same associations are found for the facet present on the coxal bone and on the sacrum. A connection between the AOVs of two bones exists. It is a contact facet between them.

4.2.2. *Pubic spine*

4.2.2.1. *Definition*

– Synonyms: None.

– Description: On the pubic ramus, close to the symphysis, a small bone spur may be present.

4.2.2.2. *Illustrations*

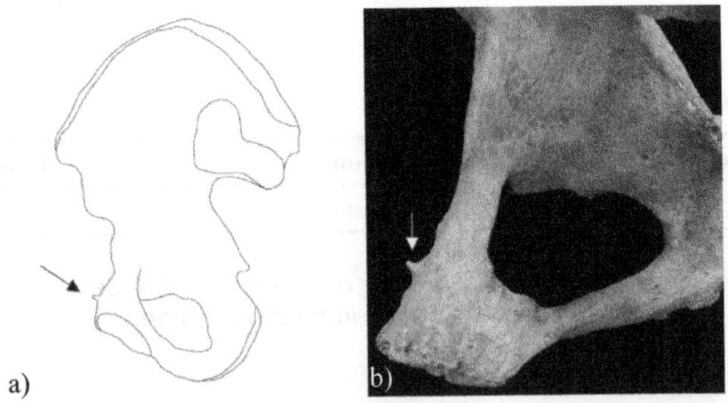

Figure 4.10. *Pubic spine: (a) diagram, anterior view; (b) dry bone, anterior view. For a color version of this figure, see www.iste.co.uk/verna/asymptomatic.zip*

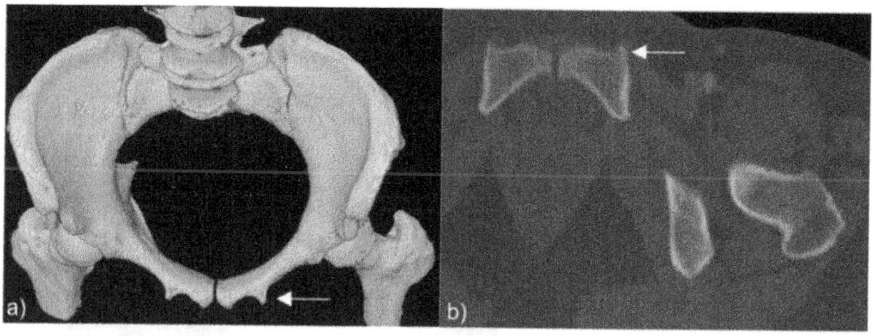

Figure 4.11. *Pubic spine: (a) 3D reconstruction, superior view; (b) DICOM image, axial section. For a color version of this figure, see www.iste.co.uk/verna/asymptomatic.zip*

4.2.2.3. *Population frequencies*

Frequency	Population	References
3.8	Mediterranean	(Verna 2014)

Table 4.11. *Frequency of the pubic spine in medical imaging*

Frequency	Population	References
No study		

Table 4.12. *Frequency of the pubic spine on dry bone*

4.2.2.4. *Association(s)*

It is mostly present in women over 40 and on the left side.

4.2.3. *Acetabular mark*

4.2.3.1. *Definition*

– Synonym: Acetabular crease.

– Description: A groove is located on the articular surface of the acetabulum, most often in its middle part at the trace level of the suture of the three bones. Different shapes are visible: a notch, a cavity, a groove or a bulge.

4.2.3.2. *Illustrations*

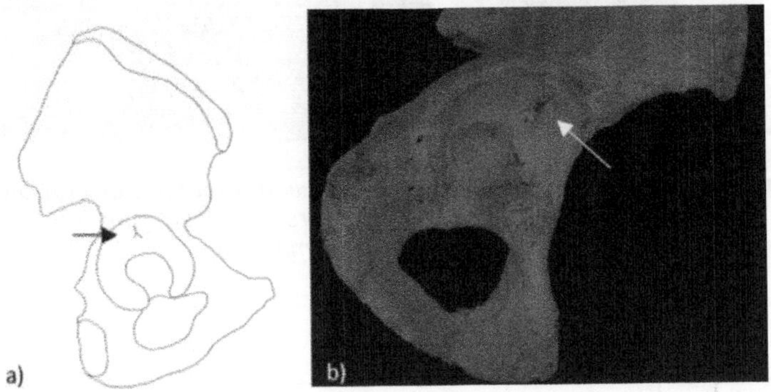

Figure 4.12. *Acetabular mark: (a) diagram, lateral view; (b) dry bone, lateral view. For a color version of this figure, see www.iste.co.uk/verna/asymptomatic.zip*

4.2.3.3. *Population frequencies*

Frequency	Population	Reference(s)
	No study	

Table 4.13. *Frequency of acetabular marks in medical imaging*

Frequency	Population	References
9.9	French (9th–12th centuries)	(Verna 2014)
11.8	French (4th–17th centuries)	(Mafart 2005)
12.5	French (4th–6th centuries)	(Castex 1994)
17.9	French (19th–20th centuries)	(Verna 2014)
19.6	American (17th–18th centuries)	(Saunders 1978)
25.4	American (Eskimo, 16th–18th centuries)	(Saunders 1978)
33.5	French (4th–5th centuries)	(Castex 1994)
34	French (Neolithic)	(Thomas 2011)
44.3	American (8th–10th centuries)	(Saunders 1978)

Table 4.14. *Frequency of acetabular marks on dry bone*

4.2.3.4. *Association(s)*

No association was found for this AOV.

5

The Lower Limb

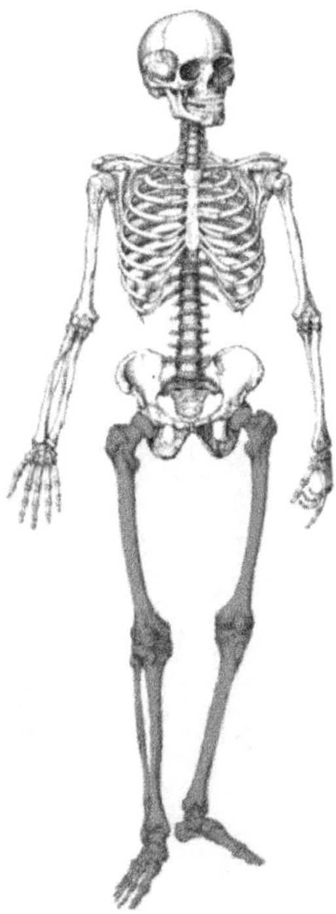

Figure 5.1. *Human skeleton. The lower limb is presented in red. For a color version of this figure, see www.iste.co.uk/verna/asymptomatic.zip*

In this chapter, the 24 AOVs present on the lower limb are described. In the foot, only the calcaneus, talus and supernumerary bones are described. There are very few variations on the rest of the foot, and these elements will not be discussed in this book.

5.1. Femur

The femur consists of three epiphyses proximally: the femoral head, greater trochanter and lesser trochanter. The distal epiphysis is composed of a secondary ossification center (the femoral condyles). Fusion of the epiphyses occurs around 14–20 years of age (Scheuer and Black 2004; Scheuer et al. 2009).

In terms of the femur, seven AOVs are observable and all of them from the earliest age.

5.1.1. *Supracondylar apophysis*

5.1.1.1. *Definition*

– Synonyms: None.

– Description: It includes a bony spine located on the diaphysis above the medial or lateral femoral condyle.

5.1.1.2. *Illustrations*

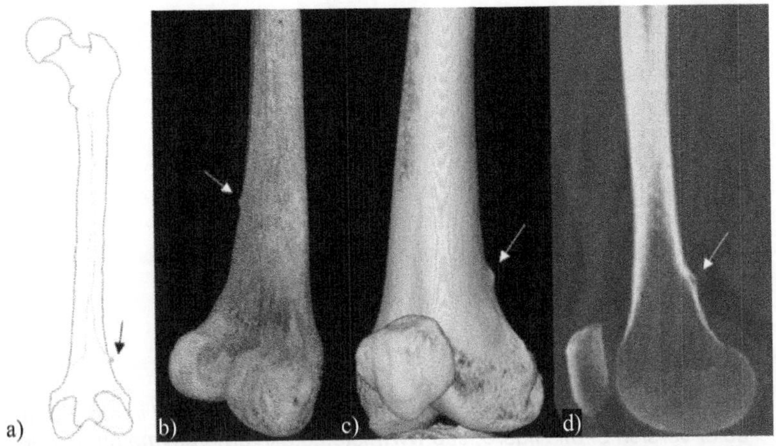

Figure 5.2. *Supracondylar apophysis: (a) diagram, right femur, posterior view; (b) dry bone, left femur, posteromedial view; (c) 3D reconstruction, left femur, anterior view; (d) DICOM image, left femur, sagittal section. For a color version of this figure, see www.iste.co.uk/verna/asymptomatic.zip*

5.1.1.3. *Population frequencies*

Apophysis	Frequency	Population	References
Medial	0.1	Mediterranean	(Verna 2014)
Lateral	1.2	Mediterranean	(Verna 2014)

Table 5.1. *Frequency of supracondylar apophysis in medical imaging*

Apophysis	Frequency	Population	References
Medial	0	French (9th–12th centuries)	(Verna 2014)
	1	French (19th–20th centuries)	(Verna 2014)
Lateral	0	French (19th–20th centuries)	(Verna 2014)
	1.2	French (9th–12th centuries)	(Verna 2014)

Table 5.2. *Frequency of supracondylar apophysis on dry bone*

5.1.1.4. *Association(s)*

No association was found for this AOV.

5.1.2. *Fabella*

5.1.2.1. *Definition*

– Synonyms: None.

– Description: It includes a small accessory bone, the size of the sesamoids, located posterior to the lateral condyle of the femur and integrated with the tendon of the gastrocnemius muscle (Duncan and Dahm 2003).

5.1.2.2. *Illustrations*

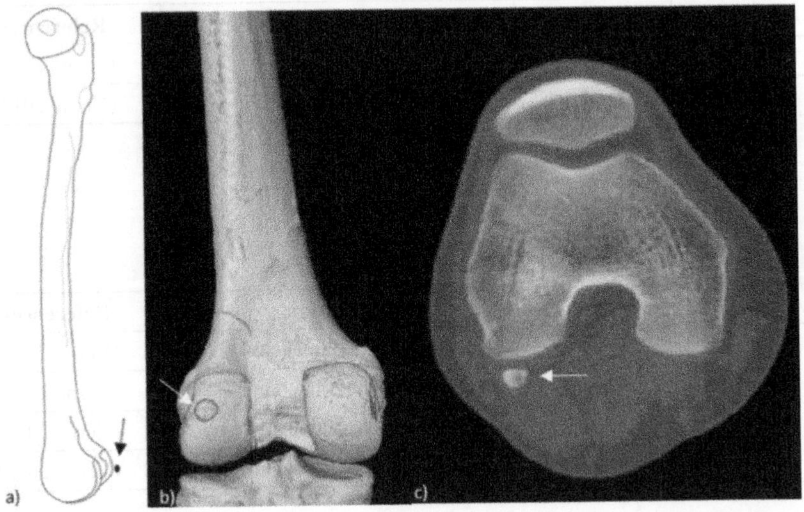

Figure 5.3. *Fabella: (a) diagram, right femur, posteromedial view; (b) 3D reconstruction, left femur, posterior view; (c) DICOM image, left femur, axial section. For a color version of this figure, see www.iste.co.uk/verna/asymptomatic.zip*

5.1.2.3. *Population frequencies*

Frequency	Population	References
10	American	(Johnson and Brogdon 1982)
19.9	Mediterranean	(Verna 2014)
22.8	Turkish	(Egerci et al. 2017)

Table 5.3. *Frequency of the fabella in medical imaging*

Frequency	Population	References
34.7	Japanese (21st century)	(Kawashima et al. 2007)

Table 5.4. *Frequency of the fabella on dry bone*

5.1.2.4. *Association(s)*

It is predominantly bilateral (Johnson and Brogdon 1982) and more present on the left when unilateral.

5.1.3. *Third trochanter*

5.1.3.1. *Definition*

– Synonyms: None.

– Description: It includes a prominent structure located under the greater trochanter in the upper part of the gluteal tuberosity as a bony prominence or tuberculus.

5.1.3.2. *Illustrations*

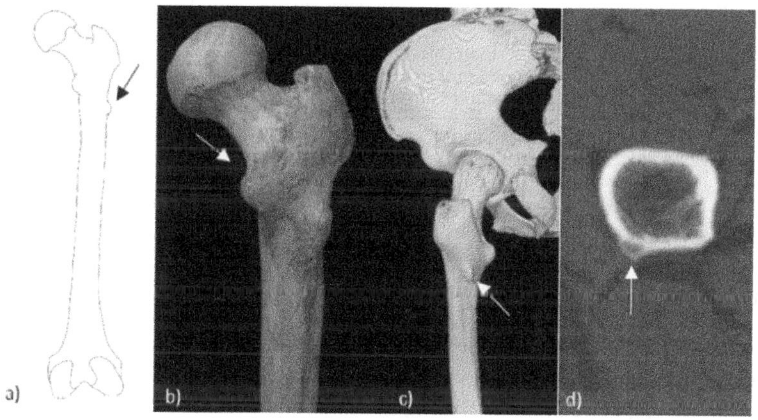

Figure 5.4. *Third trochanter: (a) diagram, right femur, posterior view; (b) dry bone, right femur, posterior view; (c) 3D reconstruction, left femur, posterolateral view; (d) DICOM image, left femur, axial section. For a color version of this figure, see www.iste.co.uk/verna/asymptomatic.zip*

5.1.3.3. *Population frequencies*

Frequency	Population	References
4.3	Mediterranean	(Verna 2014)

Table 5.5. *Frequency of the third trochanter in medical imaging*

Frequency	Population	References
3	French (Neolithic)	(Thomas 2011)
3.3	Polish (13th–17th centuries)	(Bolanowski et al. 2005)
5	French (1st–5th centuries)	(Murail 1996)
6.2	Polish (13th–14th centuries)	(Bolanowski et al. 2005)
7	American (Eskimo, 16th–18th centuries)	(Saunders 1978)
8	Polish (17th–18th centuries)	(Bolanowski et al. 2005)
13.8	French (9th–12th centuries)	(Verna 2014)
16.6–33	French (4th–12th centuries)	(Castex 1994)
17.5	French (19th–20th centuries)	(Verna 2014)
23.1	French (4th–17th centuries)	(Debono 2006)
29.5	French (7th–14th centuries)	(Chrismant 2011)
41.3	American (8th–10th centuries)	(Saunders 1978)
41.7	American (17th–18th centuries)	(Saunders 1978)
50	French (8th–12th centuries)	(Crubézy 1988)

Table 5.6. *Frequency of the third trochanter on dry bone*

5.1.3.4. *Association(s)*

Studies show great disparities, some highlighting a higher frequency in men (Castex 1994; Murail 1996), and others in women (Castex 1994; Debono 2006; Chrismant 2011), depending on the population.

Similarly for age, the general tendency is in favor of a higher frequency in those under 30, but Castex (1994) finds a frequency that increases with age. Bilaterality also seems to be favored (Castex 1994).

For this AOV, a hypothesis is put forward: the presence of the latter could be related to the robustness of the bone. The more slender the bone, the more the third trochanter is present.

Does the bony protrusion just stand out more in view of the bone's slenderness?

5.1.4. *Hypotrochanteric fossa*

5.1.4.1. *Definition*

– Synonym: *Fossa hypotrochanterica.*

– Latin translation: *Fossa hypotrochanterica.*

– Description: It is located in the posterior-superior part of the femoral shaft, between the gluteal crest and the lateral junction, outside the insertion of the gluteus maximus muscle (Hrdlicka 1934). It has a rough and irregular bottom.

5.1.4.2. *Illustrations*

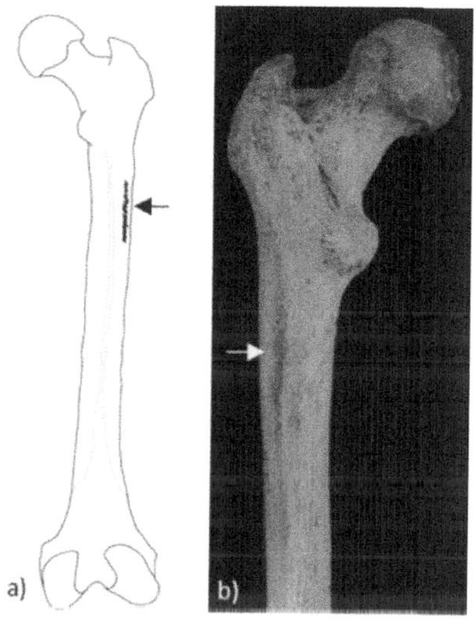

Figure 5.5. *Hypotrochanteric fossa: (a) diagram, right femur, posterior view; (b) dry bone left femur, posterior view. For a color version of this figure, see www.iste.co.uk/verna/asymptomatic.zip*

5.1.4.3. *Population frequencies*

Frequency	Population	References
No study		

Table 5.7. *Frequency of the hypotrochanteric fossa in medical imaging*

Frequency	Population	References
8.6	American (17th–18th centuries)	(Saunders 1978)
8.9	French (4th–5th centuries)	(Castex 1994)
13.4	French (19th–20th centuries)	(Verna 2014)
17.4	French (4th–17th centuries)	(Debono 2006)
20	French (4th–6th centuries)	(Castex 1994)
29.8	American (Eskimo, 6th–18th centuries)	(Saunders 1978)
30.6	American (8th–10th centuries)	(Saunders 1978)
32.6	French (12th century)	(Castex 1994)
33.1	French (9th–12th centuries)	(Verna 2014)
38.9	French (5th–7th centuries)	(Castex 1994)
53.3	French (8th–12th centuries)	(Crubézy 1988)
53.7	French (5th–6th centuries)	(Castex 1994)

Table 5.8. *Frequency of the hypotrochanteric fossa on dry bone*

5.1.4.4. *Association(s)*

Hypotrochanteric fossa is more bilateral (Saunders 1978; Crubézy 1988; Castex 1994; Debono 2006). The majority of studies show a higher frequency in people under 30 years old.

5.1.5. *Allen's fossa*

5.1.5.1. *Definition*

– Synonyms: Its synonyms are print of Berteaux, depression of the anterior femoral neck, anterior cervical imprint and fossa of Allen.

– Description: It is located in the lower half of the anterior aspect of the femoral neck and is defined as a small depression with loss of cortical bone and appearance of cancellous bone trabeculation.

There are two hypotheses for its origin: an ossification disorder (Crubézy 1988) or extension of the leg and rotation of the femoral head in the acetabulum resulting in intensive stress on the iliofemoral ligament (Odgers 1931; Capasso et al. 1999).

5.1.5.2. *Illustrations*

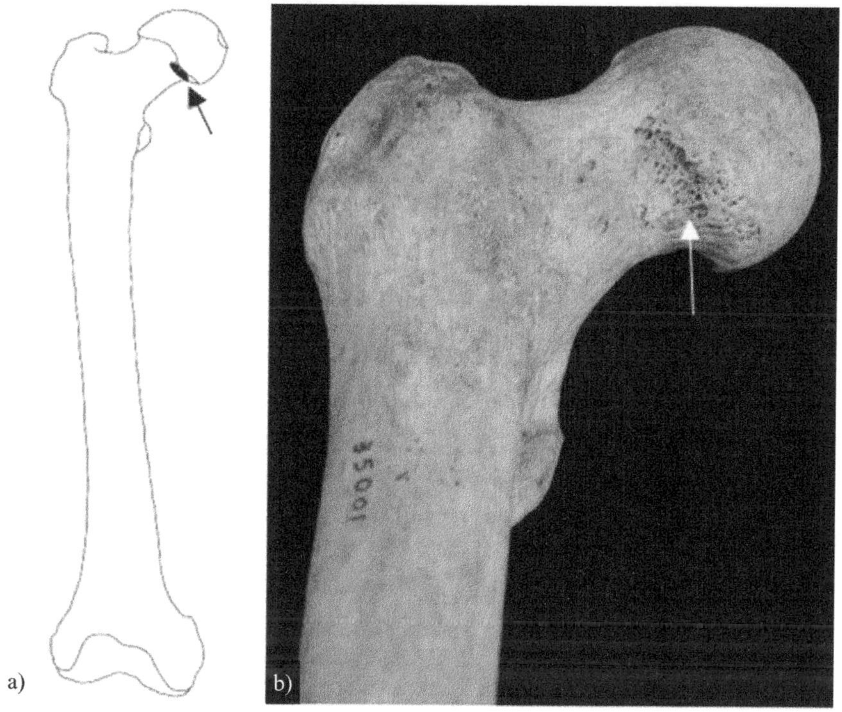

Figure 5.6. *Allen's fossa: (a) diagram, right femur, anterior view; (b) dry bone, right femur, anterior view. For a color version of this figure, see www.iste.co.uk/verna/asymptomatic.zip*

5.1.5.3. *Population frequencies*

Frequency	Population	References
No study		

Table 5.9. *Frequency of Allen's fossa in medical imaging*

Frequency	Population	References
3.3	French (19th–20th centuries)	(Verna 2014)
5.8	French (5th–6th centuries)	(Castex 1994)
9	French (Neolithic)	(Thomas 2011)
15.6	French (9th–12th centuries)	(Verna 2014)
21.5	French (4th–5th centuries)	(Castex 1994)
26.5	French (4th–6th centuries)	(Castex 1994)
26.5	French (4th–17th centuries)	(Debono 2006)
38.9	French (8th–12th centuries)	(Crubézy 1988)
41	Nigerian (20th century)	(Kostick 1963)
46	Italian (20th century)	(Radi et al. 2013)
47.5	French (5th–7th centuries)	(Castex 1994)
50	French (12th century)	(Castex 1994)

Table 5.10. *Frequency of Allen's fossaon dry bone*

5.1.5.4. *Association(s)*

Allen's fossa is more common in those under 30 years of age (Kostick 1963; Crubézy 1988; Castex 1994; Debono 2006; Thomas 2011). This fossa does not show a sex association, except in the study by Castex (1994) where women were more affected.

5.1.6. *Posterior cervical (acetabular) imprint*

5.1.6.1. *Definition*

– Synonyms: Posterior cervical (acetabular) plate and plaque.

– Description: "A rough plate, located near the anterosuperior edge of the facet joint, without continuity with the latter" (Mafart 1984).

Hypothesis for its origin is frequent squatting (Kostick 1963; Kennedy 1989).

5.1.6.2. *Illustrations*

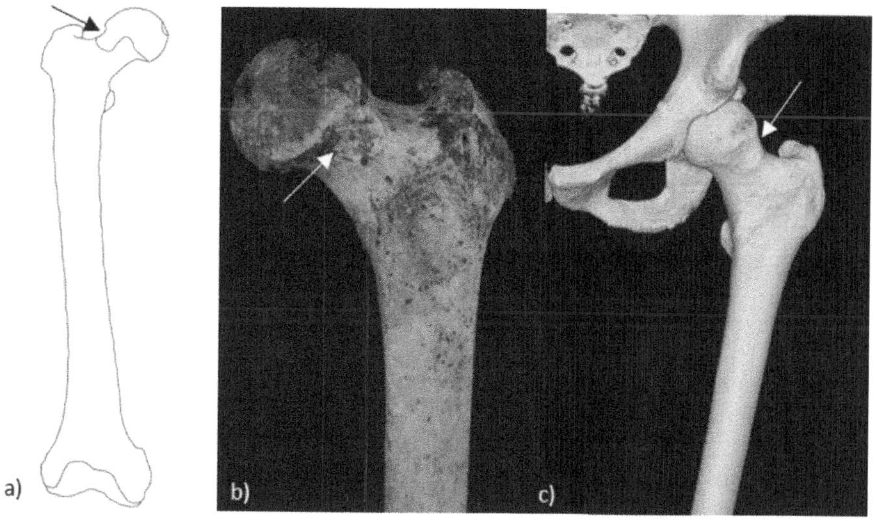

Figure 5.7. *Posterior cervical (acetabular) imprint: (a) diagram, right femur, anterior view; (b) dry bone, left femur, anterior view; (c) 3D reconstruction, left femur, anterior view. For a color version of this figure, see www.iste.co.uk/verna/asymptomatic.zip*

5.1.6.3. *Population frequencies*

The very low frequency obtained for imaging compared to dry bone can be explained by an under-scoring of this AOV due to the reconstruction, which smoothes the bone surface and which must have decreased the possibility of correctly visualizing the latter.

Frequency	Population	References
9.7	Mediterranean	(Verna 2014)

Table 5.11. *Frequency of posterior cervical (acetabular) imprint in medical imaging*

Frequency	Population	References
31	French (4th–17th centuries)	(Debono 2006)
38.1	French (9th–12th centuries)	(Verna 2014)
66.7	French (19th–20th centuries)	(Verna 2014)
67.8	French (8th–12th centuries)	(Crubézy 1988)
87	Italian (20th century)	(Radi et al. 2013)

Table 5.12. *Frequency of posterior cervical (acetabular) imprint on dry bone*

5.1.6.4. *Association(s)*

The impression is more present in men and bilaterally. However, when it is unilateral, the left side predominates. The frequency of this AOV increases with age.

5.1.7. *Poirier's facet*

5.1.7.1. *Definition*

– Synonym: Anterior cartilage excursion.

– Latin translation: *Eminentia articularis colli femoris.*

– Description: It occurs due to extension of the femoral head's articular surface at the anterior surface of the neck.

Hypothesis for its origin is hyperflexion of the hip resulting from repeated squatting (Capasso et al. 1999).

Caution: This facet can be easily confused with posterior cervical (acetabular) imprint (22% intraobserver error) (Saunders 1978). In Verna's (2014) study, this is coupled with posterior cervical (acetabular) imprint during rating.

5.1.7.2. *Illustration*

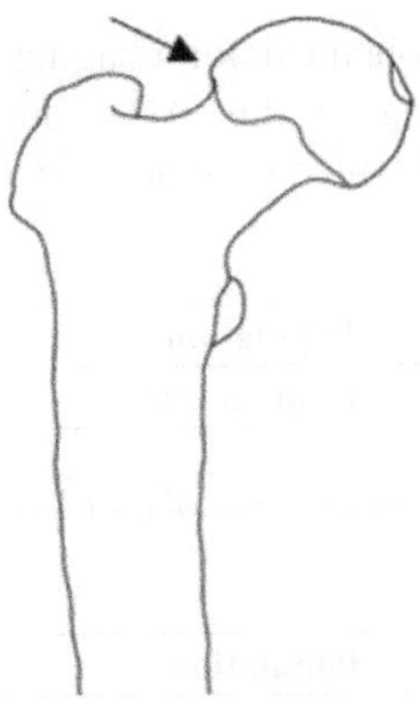

Figure 5.8. *Poirier's facet: diagram, right femur, anterior view*

5.1.7.3. *Population frequencies*

Frequency	Population	References
	No study	

Table 5.13. *Frequency of Poirier's facet in medical imaging*

Frequency	Population	References
3	Italian (20th century)	(Radi 2013)
3.9	French (5th–6th centuries)	(Castex 1994)
4.1	French (4th–17th centuries)	(Debono 2006)
11	French (4th–5th centuries)	(Castex 1994)
13	French (12th century)	(Castex 1994)
19	French (5th–7th centuries)	(Castex 1994)
31.4	French (4th–6th centuries)	(Castex 1994)
56	Nigerian (20th century)	(Kostick 1963)

Table 5.14. *Frequency of Poirier's facet on dry bone*

5.1.7.4. *Association(s)*

The frequency is greater in men than in women. Bilaterality is more often observed than unilaterality for this variation.

5.2. Patella

The patella is formed from multifocal bone centers that appear after birth (1–3 years). At puberty, the patella has its adult appearance.

A fusion failure between the different ossification centers can result in the formation of a bipartite or even multipartite patella (Scheuer and Black 2004; Scheuer et al. 2009).

The patella has three AOVs. As soon as the bone centers appear, the AOVs can be sized.

5.2.1. *Dorsal defect*

5.2.1.1. *Definition*

– Synonyms: None.

– Description: It is located in the superior-lateral part of the patella, a round depression with sclerotic edges.

This is due to an ossification defect (Jonhson 1982) or to repeated mechanical stress (Lawson 1994).

5.2.1.2. *Illustrations*

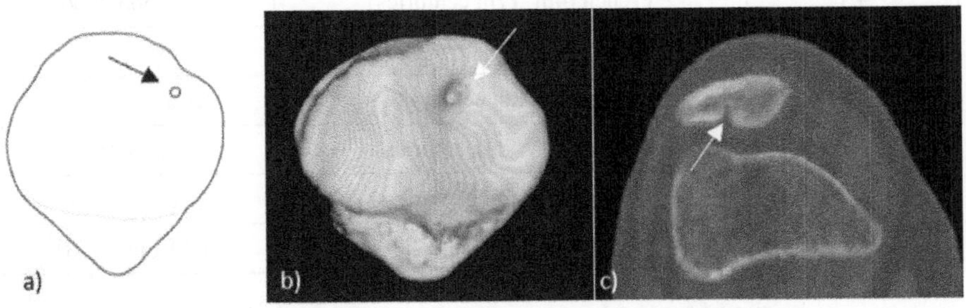

Figure 5.9. *Dorsal defect: (a) diagram, right patella, posterior view; (b) 3D reconstruction, right patella, posterior view; (c) DICOM image, right patella, axial section. For a color version of this figure, see www.iste.co.uk/verna/asymptomatic.zip*

5.2.1.3. *Population frequencies*

Frequency	Population	References
1	Mediterranean	(Verna 2014)
1	American	(Johnson and Brogdon 1982)

Table 5.15. *Frequency of dorsal defects in medical imaging*

Frequency	Population	References
0	French (19th–20th centuries)	(Verna 2014)

Table 5.16. *Frequency of dorsal defects on dry bone*

5.2.1.4. *Association(s)*

There is no link except for sex. Women demonstrate a higher frequency (Johnson and Brogdon 1982; Lawson 1994).

5.2.2. *Bipartite patella*

5.2.2.1. *Definition*

– Synonym: Emarginate patella.

– Description: It includes incomplete fusion of the second ossification center resulting in a "fracture line" with porous edges located at the superior-lateral margin.

There are two hypotheses for its origin: one congenital and the other of mechanical origin (repeated microtrauma to the vastus lateralis muscle) (O'Brien et al. 2011).

5.2.2.2. *Illustrations*

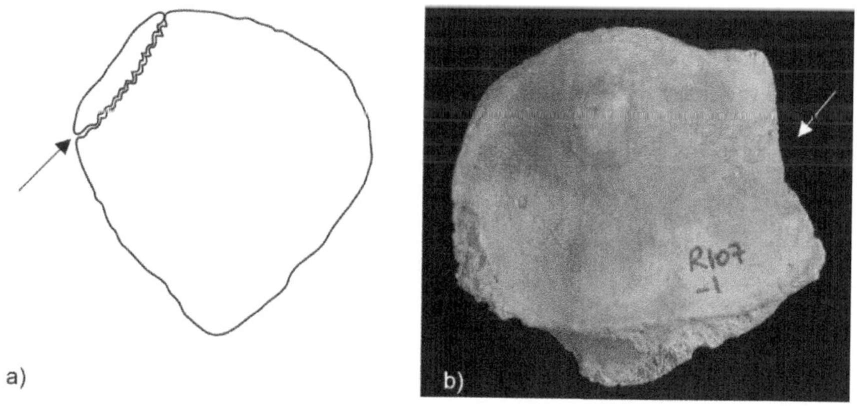

Figure 5.10. *Bipartite patella: (a) diagram, right patella, anterior view; (b) dry bone, right patella, posterior view. For a color version of this figure, see www.iste.co.uk/verna/asymptomatic.zip*

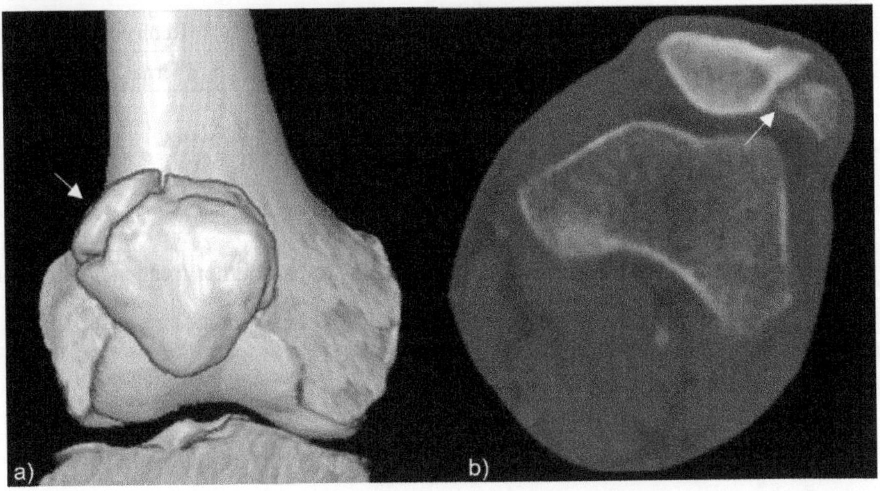

Figure 5.11. *Bipartite patella: (a) 3D reconstruction, right patella, anterior view; (b) DICOM image, right patella, axial section. For a color version of this figure, see www.iste.co.uk/verna/asymptomatic.zip*

5.2.2.3. *Population frequencies*

Frequency	Population	References
1.2	Turkish	(Kose et al. 2015)
1.8	Turkish	(Duran et al. 2016)
1.9	Mediterranean	(Verna 2014)

Table 5.17. *Frequency of bipartite patella in medical imaging*

Frequency	Population	References
0	American (17th–18th centuries)	(Saunders 1978)
0	American (8th–10th centuries)	(Saunders 1978)
1.4	English (13th–16th centuries)	(Anderson 2002)
2.5	French (4th–17th centuries)	(Debono 2006)
6.3	American (Eskimo, 16th–18th centuries)	(Saunders 1978)
8.5	French (19th–20th centuries)	(Verna 2014)

Table 5.18. *Frequency of bipartite patella on dry bone*

5.2.2.4. *Association(s)*

It is more present in men.

5.2.3. *Vastus notch*

5.2.3.1. *Definition*

– Synonyms: None.

– Description: It includes a highly developed notch with a smooth edge located at the superior-lateral angle of the patella.

There are two hypotheses for its origin: one congenital (Angel 1987) and the other mechanical (vascularization defect) (Capasso et al. 1999).

5.2.3.2. *Illustrations*

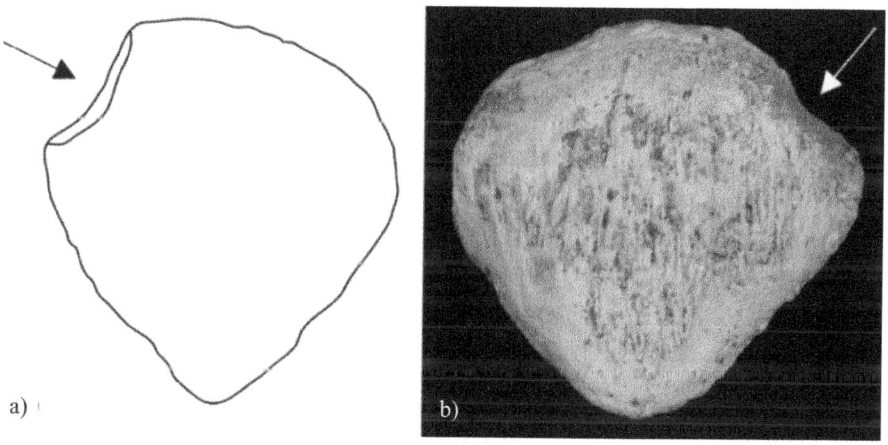

Figure 5.12. *Vastus notch: (a) diagram, right patella, anterior view; (b) dry bone, left patella, anterior view. For a color version of this figure, see www.iste.co.uk/verna/asymptomatic.zip*

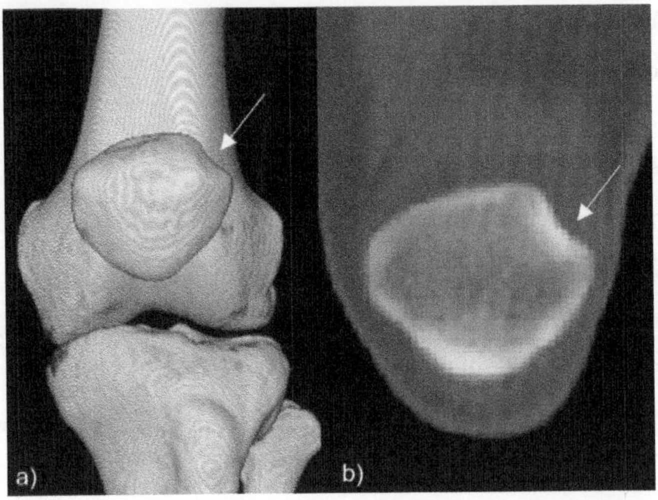

Figure 5.13. *Vastus notch: (a) 3D reconstruction, left patella, anterior view; (b) DICOM image, left patella, coronal section. For a color version of this figure, see www.iste.co.uk/verna/asymptomatic.zip*

5.2.3.3. *Population frequencies*

Frequency	Population	References
15	Mediterranean	(Verna 2014)

Table 5.19. *Frequency of vastus notch in medical imaging*

Frequency	Population	References
10	French (4th–5th centuries)	(Castex 1994)
10	French (1st–5th centuries)	(Murail 1996)
14.8	American (8th–10th centuries)	(Saunders 1978)
22.2	French (4th–6th centuries)	(Castex 1994)
24.4	American (17th–18th centuries)	(Saunders 1978)
24.6	French (4th–17th centuries)	(Debono 2006)
25.6	French (5th–7th centuries)	(Castex 1994)
34	French (Neolithic)	(Thomas 2011)
34.8	American (Eskimo, 16th–18th centuries)	(Saunders 1978)
34.9	French (19th–20th centuries)	(Verna 2014)

Table 5.20. *Frequency of vastus notch on dry bone*

5.2.3.4. *Association(s)*

Some previous studies (Murail 1996; Thomas 2011) show a higher frequency in men and bilaterally.

5.3. Tibia

The tibia forms from a primary ossification center constituting the diaphysis, and then secondary ossification centers, appearing after birth, form the epiphyses. They fuse around 14–20 years of age (Scheuer and Black 2004; Scheuer et al. 2009). The two AOVs observed on the tibia are observed from the earliest age.

5.3.1. *Lateral squatting facet*

5.3.1.1. *Definition*

– Synonyms: None.

– Description: It includes extension of the distal articular surface of the tibia forming a continuous depression located between the anterior junction of the tibia's distal epiphysis and the neck of the talus.

It occurs as a result of frequent and prolonged pressure when bending the leg over the foot (especially when squatting) (Boulle 2001).

5.3.1.2. *Illustrations*

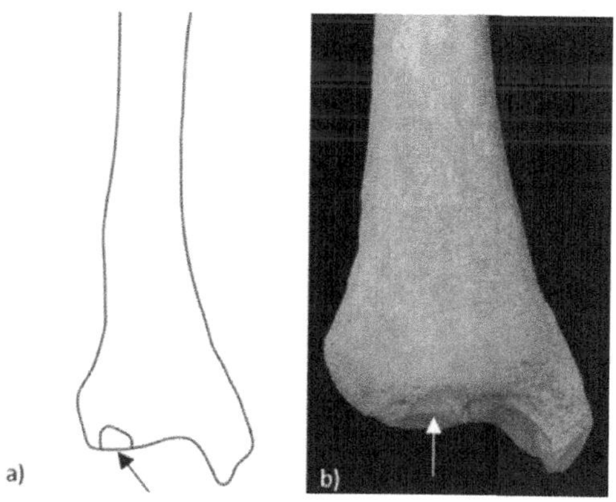

Figure 5.14. *Lateral squatting facet: (a) diagram, right tibia, anterior view; (b) dry bone, right tibia, anterior view. For a color version of this figure, see www.iste.co.uk/verna/asymptomatic.zip*

5.3.1.3. *Population frequencies*

The differences in frequency observed between populations can be explained by the origin and practice of these populations and in particular by a more or less used squatting position (Boulle 2001).

Frequency	Population	References
No study		

Table 5.21. *Frequency of lateral squatting facet in medical imaging*

Frequency	Population	References
31.1	French (5th–6th centuries)	(Castex 1994)
39.3	French (19th–20th centuries)	(Verna 2014)
41	French (4th–5th centuries)	(Castex 1994)
42	French (4th–6th centuries)	(Castex 1994)
59	French (12th century)	(Castex 1994)
60.2	French (5th–7th centuries)	(Castex 1994)
63.9	French (9th–12th centuries)	(Verna 2014)
77.4	Indian (20th century)	(Singh 1959)

Table 5.22. *Frequency of lateral squatting facet on dry bone*

5.3.1.4. *Association(s)*

Lateral squatting facets are mostly seen in women and bilaterally.

5.3.2. *Medial squatting facet*

5.3.2.1. *Definition*

– Synonyms: None.

– Description: It includes an extension of the distal articular surface of the tibia forming a depression between the anterior junction of the tibia's distal epiphysis and the neck of the talus.

It occurs as a result of frequent and prolonged pressure when bending the leg over the foot (especially when squatting) (Boulle 2001).

5.3.2.2. *Illustrations*

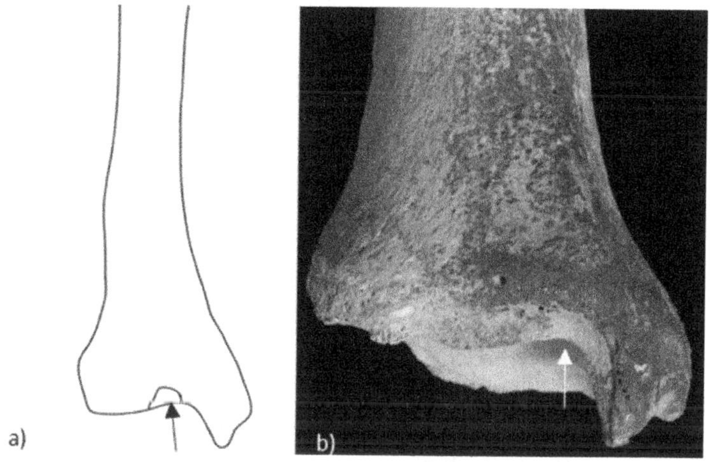

Figure 5.15. *Medial squatting facet: (a) diagram, right tibia, anterior view; (b) dry bone, right tibia, anterior view. For a color version of this figure, see www.iste.co.uk/verna/asymptomatic.zip*

5.3.2.3. *Population frequencies*

Frequency	Population	References
No study		

Table 5.23. *Frequency of medial squatting facet in medical imaging*

Frequency	Population	References
1.7	Indian (20th century)	(Singh 1959)
2.3	French (19th–20th centuries)	(Verna 2014)
6.2	French (5th–7th centuries)	(Castex 1994)
7	French (4th–6th centuries)	(Castex 1994)
7	French (4th–5th centuries)	(Castex 1994)
7	French (12th century)	(Castex 1994)
8.8	French (5th–6th centuries)	(Castex 1994)
9.5	French (9th–12th centuries)	(Verna 2014)

Table 5.24. *Frequency of medial squatting facet on dry bone*

5.3.2.4. *Association(s)*

Medial squatting facets are more often found bilaterally than unilaterally.

5.4. Talus

The talus is formed from a primary ossification center, and then secondary ossification centers appear after birth.

By the age of 20, these bones are mature (the epiphyses have fused to the rest of the bone).

The talus may consist of an additional bony center and thus result in the appearance of the trigone bone that may or may not fuse to the talar body (Scheuer and Black 2004; Scheuer et al. 2009).

The four AOVs of the talus can be observed from a young age.

5.4.1. *Lateral squatting facet*

5.4.1.1. *Definition*

– Synonyms: None.

– Description: It includes articular facet located on the superior aspect of the talus, below the trochlea.

It occurs as a result of frequent and prolonged pressure when bending the leg over the foot (especially when squatting) (Boulle 2001).

Caution: Do not confuse it with trochlear joint extension.

5.4.1.2. *Illustrations*

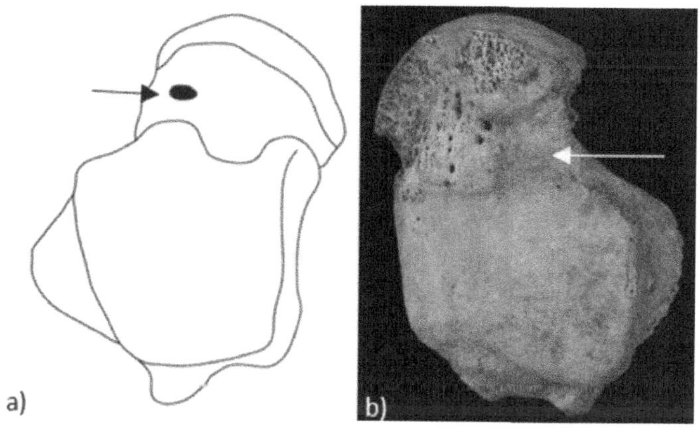

Figure 5.16. *Lateral squatting facet: (a) diagram, left talus, superior view; (b) dry bone, right talus, superior view. For a color version of this figure, see www.iste.co.uk/verna/asymptomatic.zip*

5.4.1.3. *Population frequencies*

Frequency	Population	References
No study		

Table 5.25. *Frequency of lateral squatting facet in medical imaging*

Frequency	Population	References
18–45	French (1st–18th centuries)	(Boulle 2011)
20	French (4th–6th centuries)	(Castex 1994)
21.4	French (19th–20th centuries)	(Verna 2014)
23	European (20th century)	(Barnett 1954)
25.9	French (5th–6th centuries)	(Castex 1994)
28.6	Indian (20th century)	(Singh 1959)
35	French (4th–5th centuries)	(Castex 1994)
35.9	French (9th–12th centuries)	(Verna 2014)
40	French (12th century)	(Castex 1994)
41.7	French (5th–7th centuries)	(Castex 1994)

Table 5.26. *Frequency of lateral squatting facet on dry bone*

5.4.1.4. *Association(s)*

Lateral squatting facets are mostly seen in women and bilaterally.

5.4.2. *Medial squatting facet*

5.4.2.1. *Definition*

– Synonyms: None.

– Description: Articular facet is located on the superior aspect of the talus, below the trochlea.

It occurs as a result of frequent and prolonged pressure when bending the leg over the foot (especially when squatting) (Boulle 2001).

Caution: Do not confuse it with trochlear joint extension.

5.4.2.2. *Illustrations*

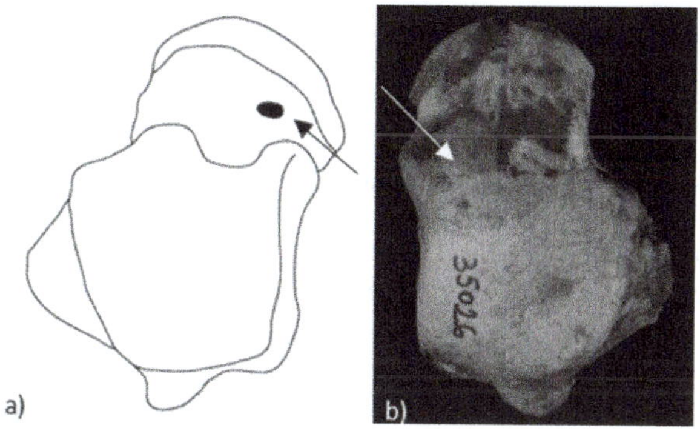

Figure 5.17. *Medial squatting facet: (a) diagram, left talus, superior view; (b) dry bone, right talus, superior view. For a color version of this figure, see www.iste.co.uk/verna/asymptomatic.zip*

5.4.2.3. *Population frequencies*

Frequency	Population	References
No study		

Table 5.27. *Frequency of medial squatting facet in medical imaging*

Frequency	Population	References
0	French (12th century)	(Castex 1994)
1.5	French (9th–12th centuries)	(Verna 2014)
4.9	French (19th–20th centuries)	(Verna 2014)
9.1	French (4th–6th centuries)	(Castex 1994)
10	French (5th–6th centuries)	(Castex 1994)
10	French (4th–5th centuries)	(Castex 1994)
10	French (Neolithic)	(Thomas 2011)
19.4	French (5th–7th centuries)	(Castex 1994)

Table 5.28. *Frequency of medial squatting facet on dry bone*

5.4.2.4. *Association(s)*

No association was found for this AOV.

5.4.3. *Os trigonum*

5.4.3.1. *Definition*

– Synonym: *Talus accessorius*.

– Latin translation: *Talus accessorius*.

– Description: Supernumerary bone located in the posterior part of the talus. When it is fused with the talus, it is termed as the Stieda process.

5.4.3.2. *Illustrations*

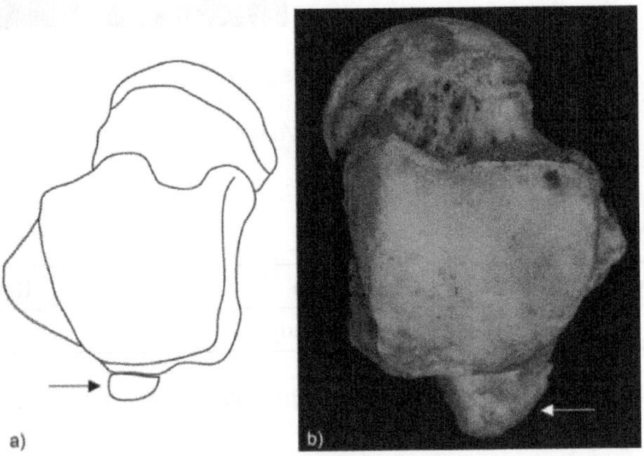

Figure 5.18. *Os trigonum: (a) diagram, left talus, superior view; (b) dry bone, right talus, superior view. For a color version of this figure, see www.iste.co.uk/verna/asymptomatic.zip*

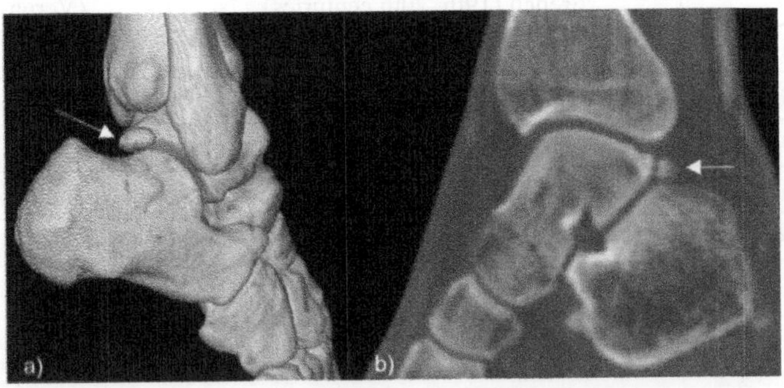

Figure 5.19. *Os trigonum: (a) 3D reconstruction, right talus, lateral view; (b) DICOM image, right talus, sagittal section. For a color version of this figure, see www.iste.co.uk/verna/asymptomatic.zip*

5.4.3.3. *Population frequencies*

On dry bone, over-rating can occur. It is indeed difficult to determine with accuracy the presence of an os trigonum, which should not be confused with a simple protuberance. In imaging, it is easier to distinguish the presence or absence of an os trigonum (especially on DICOM images).

Frequency	Population	References
2.3	Turkish	(Coskun et al. 2009)
2.9	Turkish	(Uygur et al. 2016)
19.3	Mediterranean	(Verna 2014)

Table 5.29. *Frequency of os trigonum on medical imaging*

Frequency	Population	References
1.4	American (8th–10th centuries)	(Saunders 1978)
4.5	Danish (11th–14th centuries)	(Case et al. 2017)
5.2	American (Eskimo, 16th–18th centuries)	(Saunders 1978)
6.1	English (18th–19th centuries)	(Case et al. 2017)
8.7	American (17th–18th centuries)	(Saunders 1978)
9.1	American (20th century)	(Case et al. 2017)
9.4	French (4th–17th centuries)	(Debono 2006)
13.8	French (4th–5th centuries)	(Castex 1994)
28	French (4th–6th centuries)	(Castex 1994)
42.1	French (9th–12th centuries)	(Verna 2014)

Table 5.30. *Frequency of os trigonum on dry bone*

5.4.3.4. *Association(s)*

For this variation, bilaterality predominates. In unilaterality, the left side is more often involved.

5.4.4. *Bifid inferior articular facet*

5.4.4.1. *Definition*

– Synonyms: None.

– Description: The bifid interior articular facet can be doubled and present two separate or joined articular surfaces.

5.4.4.2. *Illustrations*

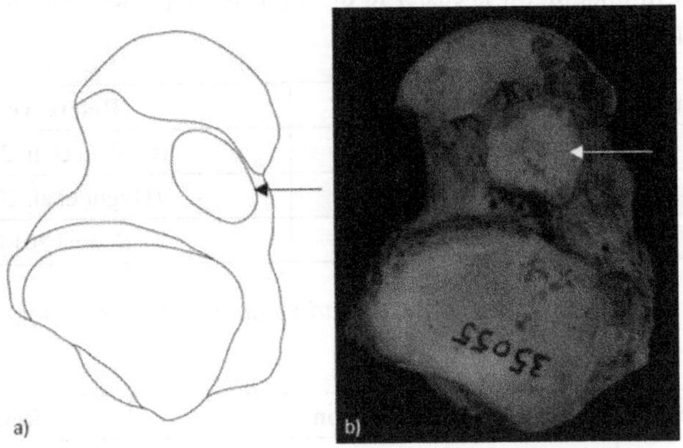

Figure 5.20. *Bifid interior articular facet: (a) diagram, right talus, inferior view; (b) dry bone, right talus, inferior view. For a color version of this figure, see www.iste.co.uk/verna/asymptomatic.zip*

5.4.4.3. *Population frequencies*

Frequency	Population	References
	No study	

Table 5.31. *Frequency of bifid interior articular facets in medical imaging*

Frequency	Population	References
13.2	French (4th–6th centuries)	(Castex 1994)
15	French	(Mafart 1984)
17	French (9th–12th centuries)	(Verna 2014)

Table 5.32. *Frequency of bifid interior articular facets on dry bone*

5.4.4.4. *Association(s)*

No association was found for this AOV.

5.5. Calcaneus

The calcaneus forms from a primary ossification center, and then secondary ossification centers appear after birth.

By the age of 20, these bones are mature (epiphyses having fused to the rest of the bone) (Scheuer and Black 2004; Scheuer et al. 2009).

The calcaneus has six AOVs. They can all be observed from a very young age.

5.5.1. *Calcaneus secondarius*

5.5.1.1. *Definition*

– Synonyms: None.

– Description: It includes a small accessory bone located at the anterior facet of the calcaneus. Often missing, it causes a notch in the anterior facet.

Hypothesis of its origin: It is a fatigue fracture as a consequence of repeated hyperflexion movements (Debono 2006).

5.5.1.2. *Illustrations*

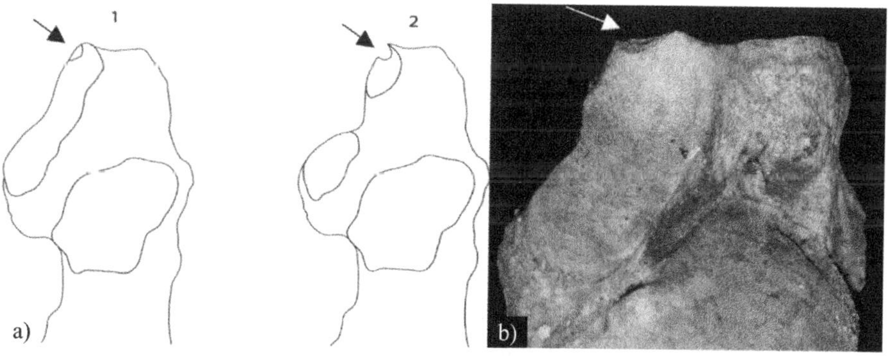

Figure 5.21. *Calcaneus secondarius: (a) fused pattern 1, unfused pattern 2, right calcaneus, superior view; (b) dry bone, right calcaneus, superior view. For a color version of this figure, see www.iste.co.uk/verna/asymptomatic.zip*

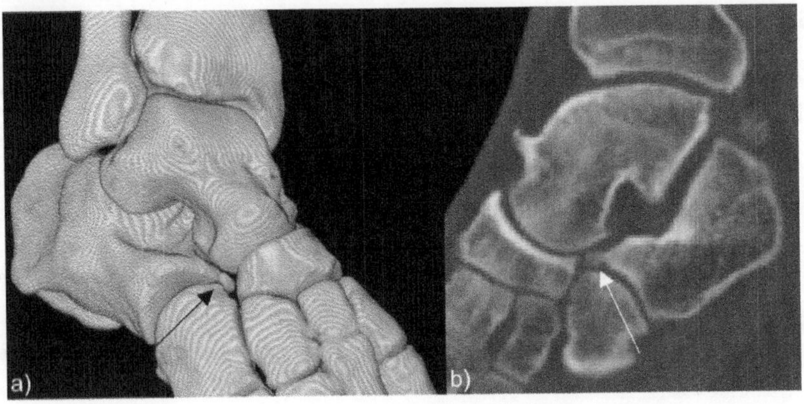

Figure 5.22. *Calcaneus secondarius: (a) 3D reconstruction, right foot, antero-lateral view; (b) DICOM image, right foot, sagittal section. For a color version of this figure, see www.iste.co.uk/verna/asymptomatic.zip*

5.5.1.3. *Population frequencies*

Frequency	Population	References
0.1	Turkish	(Uygur et al. 2016)
3.9	Mediterranean	(Verna 2014)

Table 5.33. *Frequency of calcaneus secondarius in medical imaging*

Frequency	Population	References
1	English (18th–19th centuries)	(Case et al. 2017)
1.4	American (20th century)	(Mann 1990)
2.5	American (20th century)	(Mann 1990)
3.5	Bahraini (13th–16th centuries)	(Mann 1990)
3.9	American (19th century)	(Mann 1990)
4.3	Egyptian (2000 BC)	(Mann 1990)
5.4	Norwegian (12th–17th centuries)	(Anderson 1988)
6.1	American (16th–17th centuries)	(Mann 1990)
8.8	Danish (11th–14th centuries)	(Case et al. 2017)
9.3	French (4th–17th centuries)	(Debono 2006)
14.3	French (8th–12th centuries)	(Crubézy 1988)

Table 5.34. *Frequency of calcaneus secondarius on dry bone*

5.5.1.4. *Association(s)*

It is more present in men and in a unilateral way.

5.5.2. *Absence of the anterior articular surface of the talus*

5.5.2.1. *Definition*

– Synonym: No anterior facet.

– Description: The anterior surface of the calcaneus can sometimes be completely absent.

5.5.2.2. *Illustrations*

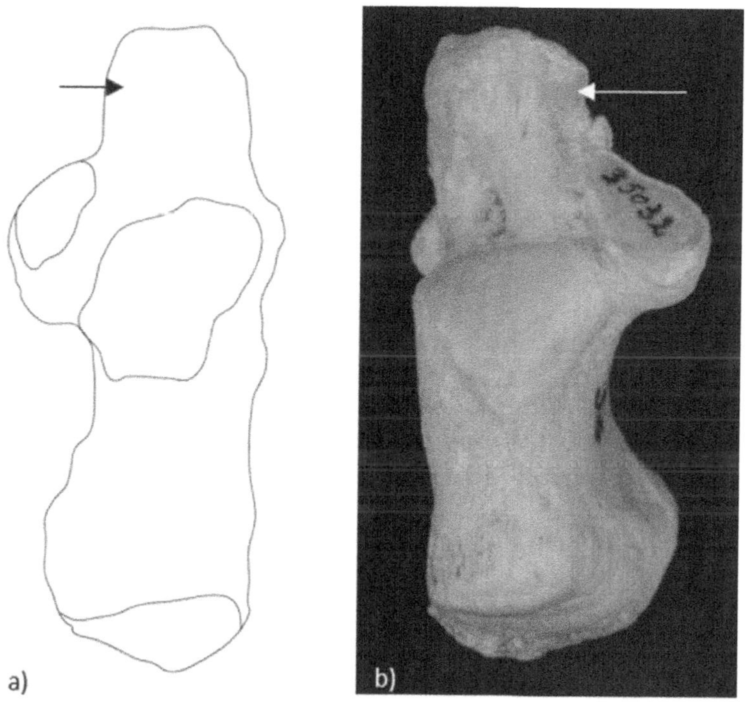

Figure 5.23. *Absence of the anterior articular surface of the talus: (a) diagram, right calcaneus, superior view; (b) dry bone, left calcaneus, superior view. For a color version of this figure, see www.iste.co.uk/verna/asymptomatic.zip*

5.5.2.3. *Population frequencies*

Frequency	Population	References
No study		

Table 5.35. *Frequency of absence of the anterior articular surface of the talus in medical imaging*

Frequency	Population	References
1.1	American (Eskimo, 16th–18th centuries)	(Saunders 1978)
1.4	American (8th–10th centuries)	(Saunders 1978)
2.9	American (17th–18th centuries)	(Saunders 1978)
3.9	French (4th–5th centuries)	(Castex 1994)
4.9	French (4th–6th centuries)	(Castex 1994)
6.1	French (9th–12th centuries)	(Verna 2014)
7.4	French (12th century)	(Castex 1994)
12	French (19th–20th centuries)	(Verna 2014)

Table 5.36. *Frequency of absence of the anterior articular surface of the talus on dry bone*

5.5.2.4. *Association(s)*

A trend for bilaterality is observed in Castex's study (1994).

5.5.3. *Bipartite anterior facet*

5.5.3.1. *Definition*

– Synonym: Double anterior facet.

– Description: The anteromedial articular surface of the calcaneus may have two distinct facets.

5.5.3.2. *Illustrations*

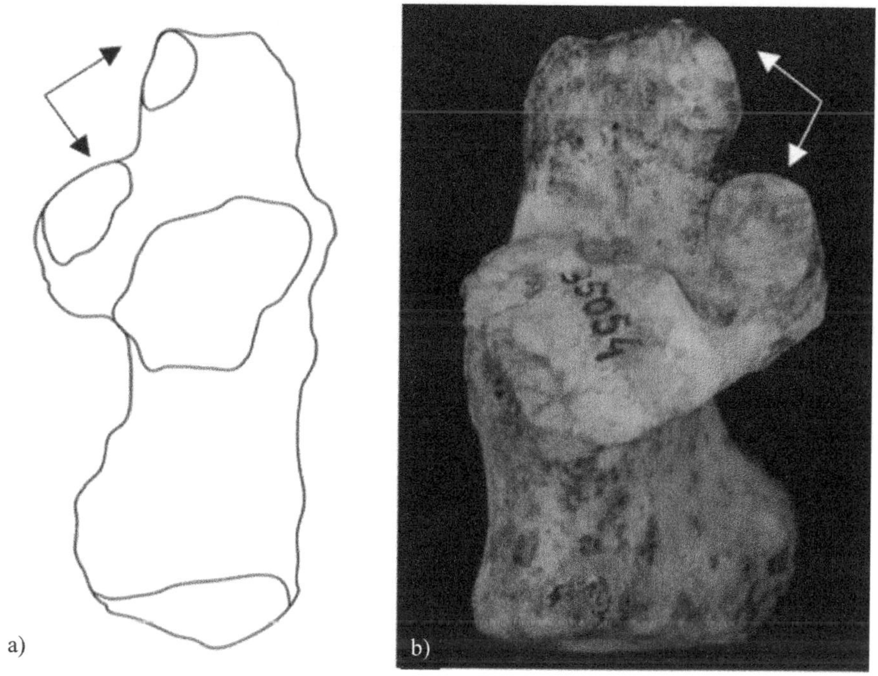

Figure 5.24. *Bipartite anterior facet: (a) diagram, right calcaneus, superior view; b) dry bone, left calcaneus, superior view. For a color version of this figure, see www.iste.co.uk/verna/asymptomatic.zip*

5.5.3.3. *Population frequencies*

Frequency	Population	References
	No study	

Table 5.37. *Frequency of the bipartite anterior facet in medical imaging*

Frequency	Population	References
7.3	American (8th–10th centuries)	(Saunders 1978)
7.8	American (Eskimo, 16th–18th centuries)	(Saunders 1978)
15	American (17th–18th centuries)	(Saunders 1978)
29.3	French (5th–7th centuries)	(Castex 1994)
35	French (4th–5th centuries)	(Castex 1994)
45	French (12th century)	(Castex 1994)
50	French (4th–6th centuries)	(Castex 1994)
51	French (Neolithic)	(Thomas 2011)
51.3	French (9th–12th centuries)	(Castex 1994)
65.9	French (19th–20th centuries)	(Verna 2014)

Table 5.38. *Frequency of the bipartite anterior facet on dry bone*

5.5.3.4. *Association(s)*

It is mostly found bilaterally.

5.5.4. *Extension of the posterior articular facet*

5.5.4.1. *Definition*

– Synonyms: None.

– Description: It includes extension of the posterior articular surface of the calcaneus.

Hypothesis of its origin: Its origin is mechanical (repeated squatting).

5.5.4.2. *Illustrations*

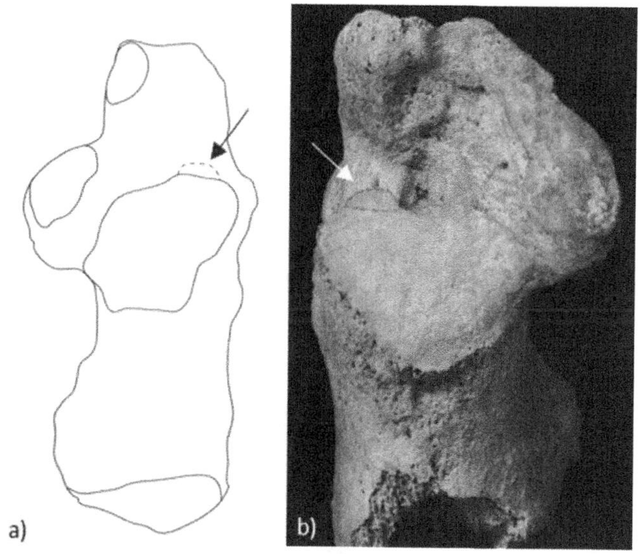

Figure 5.25. *Extension of the posterior articular facet: (a) diagram, right calcaneus, superior view; (b) dry bone, left calcaneus, superior view. For a color version of this figure, see www.iste.co.uk/verna/asymptomatic.zip*

5.5.4.3. *Population frequencies*

Frequency	Population	References
No study		

Table 5.39. *Frequency of extension of the posterior articular facet in medical imaging*

Frequency	Population(s)	Reference(s)
15.9	American (8th–10th centuries)	(Saunders 1978)
17.7	American (Eskimo, 16th–18th centuries)	(Saunders 1978)
28.4	French (4th–6th centuries)	(Castex 1994)
29	French (9th–12th centuries)	(Verna 2014)
30.1	American (17th–18th centuries)	(Saunders 1978)

Table 5.40. *Frequency of extension of the posterior articular facet on dry bone*

5.5.4.4. *Association(s)*

It is mostly found bilaterally.

5.5.5. *Peroneal trochlea*

5.5.5.1. *Definition*

– Synonyms: Fibular tubercle and peroneal tubercle.

– Description: Bony prominence located between the tendons of the long and short fibular muscles on the lateral surface of the calcaneus.

5.5.5.2. *Illustrations*

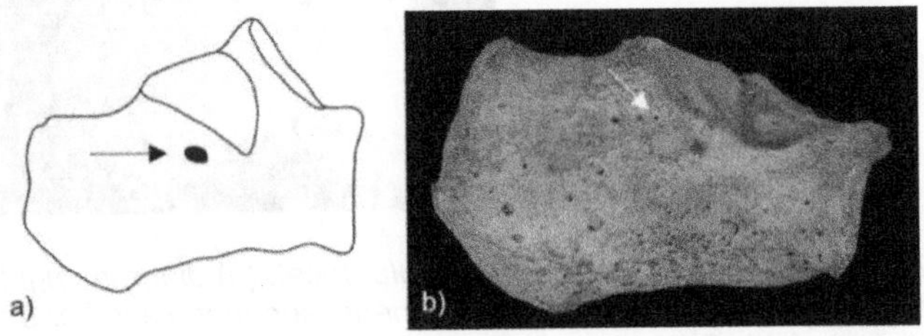

Figure 5.26. *Peroneal trochlea: (a) diagram, right calcaneus, lateral view; (b) dry bone, right calcaneus, lateral view. For a color version of this figure, see www.iste.co.uk/verna/asymptomatic.zip*

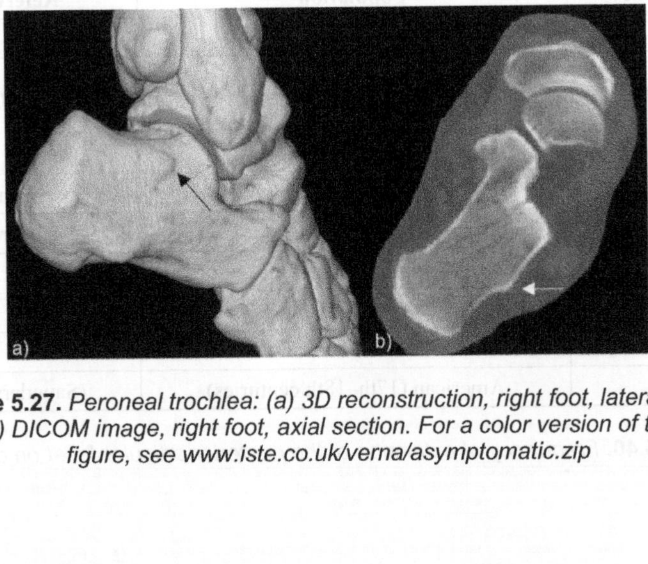

Figure 5.27. *Peroneal trochlea: (a) 3D reconstruction, right foot, lateral view; (b) DICOM image, right foot, axial section. For a color version of this figure, see www.iste.co.uk/verna/asymptomatic.zip*

5.5.5.3. *Population frequencies*

Frequency	Population	References
2.3	Mediterranean	(Verna 2014)
55	Switzerland	(Saupe et al. 2007)

Table 5.41. *Frequency of the peroneal trochlea in medical imaging*

Frequency	Population	References
14.1	French (4th–6th centuries)	(Castex 1994)
18.1	French (9th–12th centuries)	(Verna 2014)

Table 5.42. *Frequency of the peroneal trochlea on dry bone*

5.5.5.4. *Association(s)*

The peroneal trochlea is more frequent in women unilaterally and in those over 40 years of age.

5.5.6. *Retrotrochlear eminence*

5.5.6.1. *Definition*

– Synonyms: None.

– Description: It is located on the lateral surface of the bone, near the attachment of the calcaneo-fibular ligament and appears as a small bony prominence.

Hypothesis of its origin: It has a functional origin due to a repeated squatting position (Mafart 1980).

5.5.6.2. *Illustrations*

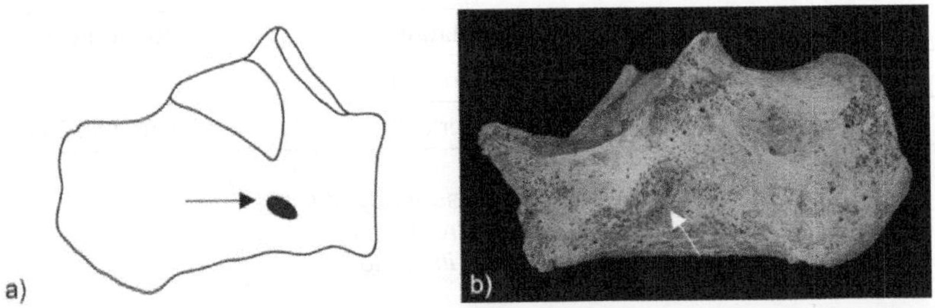

Figure 5.28. *Retrotrochlear eminence: (a) diagram, right calcaneus, lateral view; (b) dry bone, left calcaneus, lateral view. For a color version of this figure, see www.iste.co.uk/verna/asymptomatic.zip*

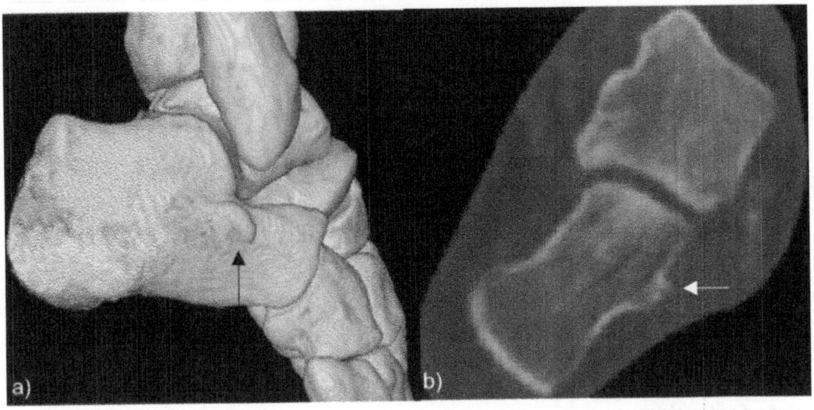

Figure 5.29. *Retrotrochlear eminence: (a) 3D reconstruction, right foot, lateral view; (b) DICOM image, right foot, axial section. For a color version of this figure, see www.iste.co.uk/verna/asymptomatic.zip*

5.5.6.3. *Population frequencies*

Frequency	Population	References
39.6	Mediterranean	(Verna 2014)
100	Switzerland	(Saupe et al. 2007)

Table 5.43. *Frequency of retrotrochlear eminence in medical imaging*

Frequency	Population	References
38.5	French (9th–12th centuries)	(Verna 2014)
73.9	French (4th–6th centuries)	(Castex 1994)

Table 5.44. *Frequency of retrotrochlear eminence on dry bone*

5.5.6.4. *Association(s)*

Retrotrochlear eminence is more frequent in women bilaterally, and in case of unilaterality the right side is more affected.

This predominance in females can be explained by a lesser robustness than in males, which could therefore allow more bony prominences to appear.

5.6. The foot

The foot has supernumerary sesamoid bones in various tarsal bones. They are presented here. These are the two most frequently found in the literature.

5.6.1. *Os tibiale*

5.6.1.1. *Definition*

– Synonyms: Accessory navicular bone, accessory navicular and accessory scaphoid.

– Latin translations: *Os naviculare, navicular secundum, os tibiale externum, naviculare secundarium.*

– Description: Accessory bone located at the posteromedial part of the navicular tuberosity.

Three types of accessory naviculars exist (Mellado et al. 2003; Offenbecker and Case 2012):

– type 1: sesamoid-sized ossicles (2–5 mm);

– type 2: persistence of an additional ossification center (>5 mm);

– type 3: fused form of type 2 (forming a horn: “horny navicular”).

5.6.1.2 *Illustrations*

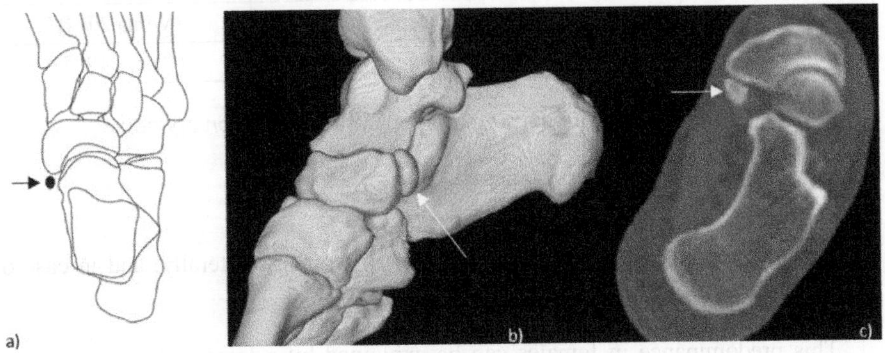

Figure 5.30. *Os tibiale: (a) diagram, right foot, superior view; (b) 3D reconstruction, right foot, medial view; (c) DICOM image, right foot, axial section. For a color version of this figure, see www.iste.co.uk/verna/asymptomatic.zip*

5.6.1.3. *Population frequencies*

The difference in frequency between the two media can be explained by a conservation problem.

This accessory bone is rarely found on dry bone and an underestimation of the frequency can then be made.

Frequency	Population	References
5.8	Turkish	(Uygur et al. 2016)
11.7	Turkish	(Coskun et al. 2009)
17.6	Greek	(Perdikakis et al. 2011)
20.2	Chinese	(Huang et al. 2014)
23.5	Mediterranean	(Verna 2014)

Table 5.45. *Frequency of the os tibiale in medical imaging*

Frequency	Population	References
0	American (17th–18th centuries)	(Saunders 1978)
0	American (8th–10th centuries)	(Saunders 1978)
0.3	American (Eskimo, 16th–18th centuries)	(Saunders 1978)
2	American (20th century)	(Offenbecker and Case 2012)
2.4	American (20th century)	(Offenbecker and Case 2012)
2.9	English (18th–19th centuries)	(Offenbecker and Case 2012)
4.2	Danish (11th–16th centuries)	(Offenbecker and Case 2012)
5	Japanese (19th–20th centuries)	(Offenbecker and Case 2012)

Table 5.46. *Frequency of the os tibiale on dry bone*

5.6.1.4. *Association(s)*

This variation is more often observed in females and in individuals older than 20. Bilaterality predominates in the lateral os tibiale.

5.6.2. *Os peroneum*

5.6.2.1. *Definition*

– Synonym: *Os fibulare.*

– Description: Round accessory bone (similar to the sesamoid) located at the calcaneocuboid junction and integrated into the tendon of the long fibular muscle (Leminor 1987).

5.6.2.2. *Illustrations*

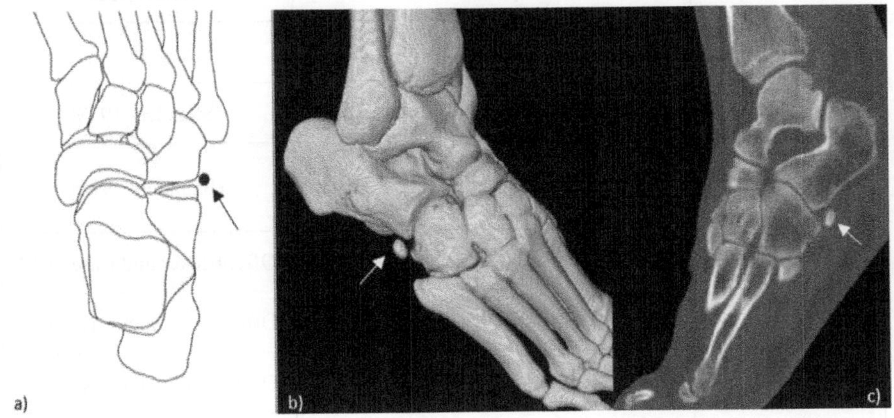

Figure 5.31. *Os peroneum: (a) diagram, right foot, superior view; (b) 3D reconstruction, right foot, lateral view; (c) DICOM image, left foot, sagittal section. For a color version of this figure, see www.iste.co.uk/verna/asymptomatic.zip*

5.6.2.3. *Population frequencies*

Frequency	Population	References
4.7	Turkish	(Coskun et al. 2009)
5.3	Turkish	(Uygur et al. 2016)
21.1	Mediterranean	(Verna 2014)

Table 5.47. *Frequency of the os peroneum in medical imaging*

Frequency	Population	References
90	South African (21st century)	(Oyedele et al. 2006)

Table 5.48. *Frequency of the os peroneum on dry bone*

5.6.2.4. *Association(s)*

Studies on this AOV are in agreement for an increase in frequency with age (a late ossification of this accessory bone). The majority of *os peroneum* bones are bilateral and no side preference has been observed.

Appendix

AOVs versus Pathologies

It is important to be familiar with osteology so as not to confuse variations with pathological or traumatic markers. For example, in a forensic setting, sternal perforation may have been mistakenly associated with a gunshot to the chest. Care should be taken to ensure that the edges of the perforation are smooth and have no discontinuity with the cortical bone. The supracondyloid process can be misjudged and treated as pathology. This is because the distal spine is in the same location, but it is oriented upward (Figure A.1). It is therefore important to be careful when sizing so as not to confuse the two types of spines.

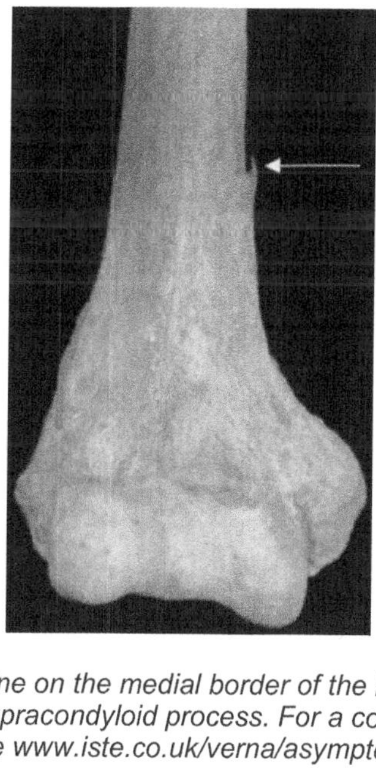

Figure A.1. *Distal spine on the medial border of the humerus, at the same location as the supracondyloid process. For a color version of this figure, see www.iste.co.uk/verna/asymptomatic.zip*

The supracondyloid tunnel can also be considered a pathological case. While it appears as several small bone bridges at the end of the distal end of the humerus, it is a pathological alteration (an exostosis).

The facet joint of the acromion with the humerus can be of pathological origin, notably with the presence of osteophytes and a porous aspect of the surface. This may be of inflammatory or traumatic origin.

This facet (Figure A.2) can be differentiated from the AOV, which has a well-defined articular surface with a smooth, regular appearance. In the absence of an associated humerus, the traumatic origin of the scapula's facet joint seems more difficult to determine (Voisin 2012).

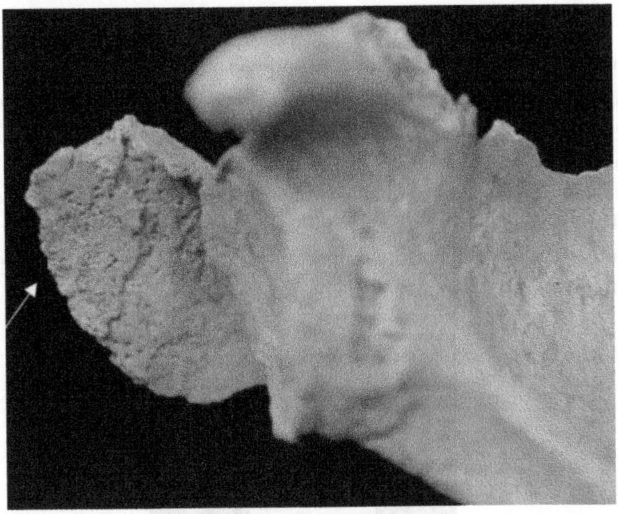

Figure A.2. *Acromial facet showing bone remodeling and porosities. For a color version of this figure, see www.iste.co.uk/verna/asymptomatic.zip*

Instead of a rib fusion, an intrathoracic bridge may also be present in the form of a neoarticular surface between two ribs. It may be traumatic in origin and follow fracture consolidation or be associated with an AOV (a joint between two ribs without the presence of fusion) (Figure A.3) (Barnes 2012).

Fusion between two ribs can also occur as a result of fracture consolidation and, in this case, is not considered an AOV (Figure A.4).

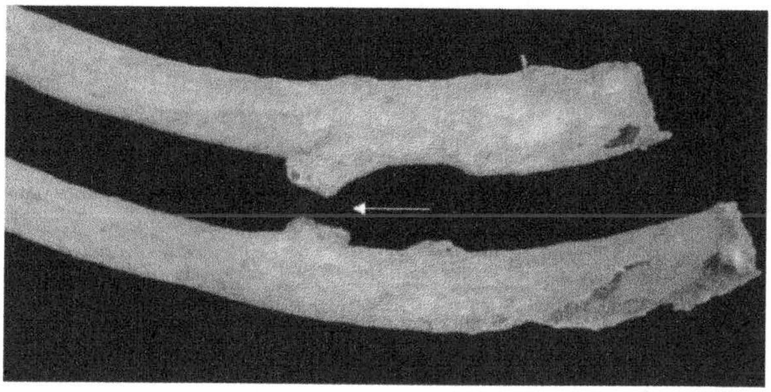

Figure A.3. *Neoarticulation between two ribs. For a color version of this figure, see www.iste.co.uk/verna/asymptomatic.zip*

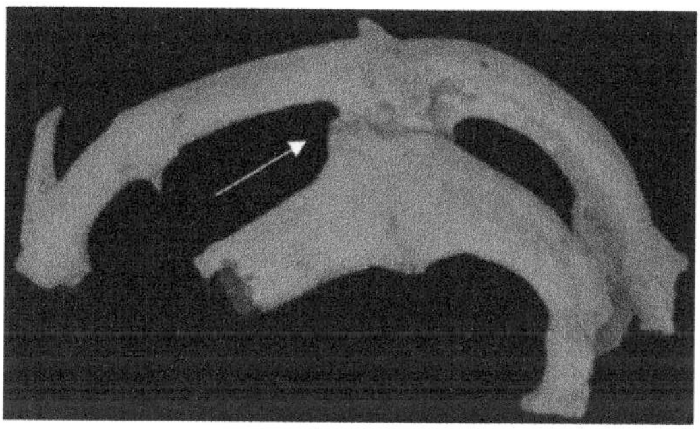

Figure A.4. *Fusion between two ribs following consolidation of a rib fracture. For a color version of this figure, see www.iste.co.uk/verna/asymptomatic.zip*

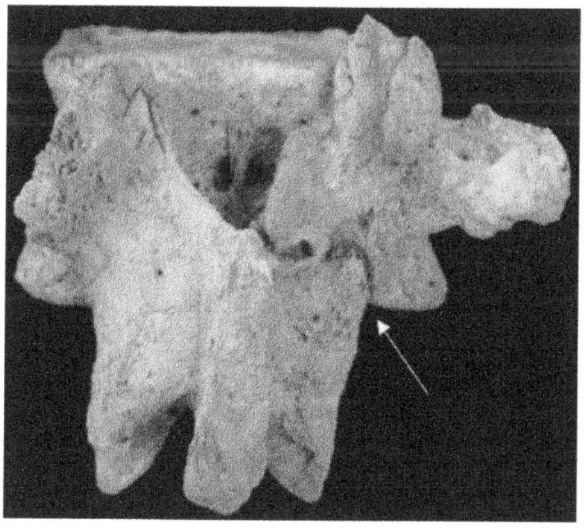

Figure A.5. *Unilateral spondylolysis of a lumbar vertebra. For a color version of this figure, see www.iste.co.uk/verna/asymptomatic.zip*

Some AOVs may take on the appearance of a fracture. For example, spondylolysis, when unilateral (Figure A.5), may be mistaken for a fracture in the spinous process. In this case, attention should be paid to the cortical bone to see if it has continuity or if cancellous bone is apparent as well as cracks around the area. All AOVs due to a fusion defect can be mistaken for a fracture.

Variations in the number of foramen can also be observed, particularly in humeral perforation or the suprascapular foramen (Voisin 2012).

The circumflex sulcus also shows variations. Some are present perpendicularly and bisect the lateral border of the scapula, while others are found below the glenoid fossa and are oblique to the lateral border of the scapula. These slight distinctions in position correspond to the same mark left by the imprint of the circumflex artery.

References

Akabori, E. (1934). Septal apertures in the humerus in Japanese, Ainu and Koreans. *Am. J. Phys. Anthropol.*, 18(3), 395–400.

Akin, K., Kosehan, D., Topcu, A., Koktener, A. (2011). Anatomic evaluation of the xiphoid process with 64-row multidetector computed tomography. *Skeletal Radiol.*, 40(4), 447–452.

Aktan, Z.A. and Savas, R. (1998). Anatomic and HRCT demonstration of midline sternal foramina. *Turkish J. Med. Sci.*, 8(5), 511–514.

Al-Motabagani, M.A. and Surendra, M. (2006). Total occipitalization of the atlas. *Anat. Sci. Int.*, 81(3), 173–180.

Albrecht, T.L., Scutter, S.D., Henneberg, M. (2007). Radiographic method to assess the prevalence of sacral spina bifida occulta. *Clin. Anat.*, 20(2), 70–174.

Altun, I. and Yuksel, K.Z. (2017). An uncommon case mimicking cervical trauma: Os odontoideum. *Turkish J. Emerg. Med.*, 17(1), 22–24.

Anderson, T. (1988). Calcaneus secundarius: An osteo-archaeological note. *Am. J. Phys. Anthropol.*, 77, 529–531.

Anderson, T. (2002). A bipartite patella in a juvenile from a medieval context. *Int. J. Osteoarchaeol.*, 12(4), 297–302.

Angel, J.L., Kelley, J.O., Parrington, M., Pinter, S. (1987). Life stresses of the free black-community as represented by the 1st-African-Baptist-Church, Philadelphia, 1823–1841. *Am. J. Phys. Anthropol.*, 74(2), 213–229.

Apaydin, M., Sarsilmaz, A., Varer, M. (2009). Third accessory (supernumerary) intrathoracic right rib. *Surg. Radiol. Anat.*, 31(8), 641–643.

Apazidis, A., Ricart, P.A., Diefenbach, C.M., Spivak, J.M. (2011). The prevalence of transitional vertebrae in the lumbar spine. *Spine J.*, 11(9), 858–862.

Ardagna, Y., Kacki, S., Verna, E., Panuel, M., Baud, M. (2014). Études paléopathologiques de cas d'une atteinte rachidienne rare : les vertèbres en papillon. *Bull. Mem. Soc. Anthropol. Paris*, 27, 72–82.

Aydinlioglu, A., Gumrukcuoglu, F.N., Koyun, N. (2010). The prevalence of supracondyloid process in the living: A radiographic study. *W. Indian Med. J.*, 59(5), 545–548.

Babinski, M.A., de Lemos, L., Babinski, M.S.D., Goncalves, M.V.T., de Paula, R.C., Fernandes, R.M.P. (2015). Frequency of sternal foramen evaluated by MDCT: A minor variation of great relevance. *Surg. Radiol. Anat.*, 37(3), 287–291.

Barnes, E. (2012). *Atlas of Developmental Field Anomalies of the Human Skeleton: A Paleopathology Perspective*. Wiley-Blackwell, Hoboken.

Barnett, C.H. (1954). Squatting facets on the European talus. *J. Anat.*, 88, 508–513.

Bayarogullari, H., Yengil, E., Davran, R., Aglagul, E., Karazincir, S., Balci, A. (2014). Evaluation of the postnatal development of the sternum and sternal variations using multidetector CT. *Diagn. Interv. Radiol.*, 20(1), 82–89.

Berry, A.C. (1974). The use of non-metrical variations of the cranium in the study of Scandinavian population movements. *Am. J. Phys. Anthropol.*, 40(3), 345–358.

Berry, A.C. (1975). Factors affecting incidence of non-metrical skeletal variants. *J. Anat.*, 120, 519–535.

Berry, A.C. and Berry, R. (1967). Epigenetic variation in the human cranium. *J. Anat.*, 101(2), 361–379.

Bilodi, A. and Gupta, S. (2005). Presence of retro transverse groove or canal in atlas vertebrae. *J. Anat. Soc. India*, 54(1), 16–18.

Bocquet-Appel, J. (1984). Biological evolution and history in 19th century Portugal. In *Multivariate Statistical Methods in Physical Anthropology*, van Vark, G.N., Howells, W.W. (eds). Springer, Dordrecht.

Bolanowski, W., Śmiszkiewicz-Skwarska, A., Polguj, M., Jędrzejewski, K.S. (2005). The occurrence of the third trochanter and its correlation to certain anthropometric parameters of the human femur. *Folia Morphol.*, 64(3), 168–175.

Boone, D., Parsons, D., Lachmann, S.M., Sherwood, T. (1985). Spina-bifida occulta – Lesion or anomaly. *Clin. Anat.*, 36(2), 159–161.

Bottosso, N. and Ghaye, B. (2008). Bifid intrathoracic rib. *J. Belge Radiol.*, 91(3), 86–87.

Boulle, E.L. (2001). Evolution of two human skeletal markers of the squatting position: A diachronic study from antiquity to the modern age. *Am. J. Phys. Anthropol.*, 115(1), 50–56.

Braga, J. (1993). Quelques exemples de développement des caractères discrets chez les grands singes. *Bull. Mem. Soc. Anthropol. Paris*, 5, 119–130.

Brasili, P., Bonfiglioli, B., Ventrella, A.R. (2002). A case of "butterfly" vertebra from Sardinia. *Int. J. Osteoarchaeol.*, 12(6), 415–419.

Brewin, A., Hill, M., Ellis, H. (2009). The prevalence of cervical ribs in a London population. *Clin. Anat.*, 22(3), 331–336.

Capasso, L., Kennedy, K.A.R., Wilczak, C.A. (1999). *Atlas of Occupational Markers on Humain Remains*. Edigraphital SpA, Teramo.

Case, D.T., Burnett, S., Nielsen, T. (2005). Os acromiale: Population differences and their etiological significance. *HOMO*, 57, 1–18.

Case, D.T., Jones, L.B., Offenbecker, A.M. (2017). Skeletal kinship analysis using developmental anomalies of the foot. *Int. J. Osteoarchaeol.*, 27(2), 192–205.

Castex, D. (1994). Mortalité, morbidité et gestion de l'espace funéraire au cours du haut Moyen Âge. Contribution spécifique de l'anthropologie biologique. PhD Thesis, Université Bordeaux 1, Bordeaux.

Chauhan, R. and Khanna, J. (2012). Absence of costal element of the foramen transversarium of atlas vertebrae. *Int. J. Res. Med. Sci.*, 1(2), 66–68.

Cheverud, J.M. and Buikstra, J.E. (1981a). Quantitative genetics of skeletal nonmetric traits in the rhesus macaques on Cayo Santiago I. Single trait heritabilities. *Am. J. Phys. Anthropol.*, 54(1), 43–49.

Cheverud, J.M. and Buikstra, J.E. (1981b). Quantitative genetics of skeletal nonmetric traits in the rhesus macaques on Cayo Santiago II. Phenotypic, genetic, and environmental correlations between traits. *Am. J. Phys. Anthropol.*, 54(1), 51–58.

Cheverud, J.M. and Buikstra, J.E. (1982). Quantitative genetics of skeletal nonmetric traits in the rhesus macaques of Cayo Santiago III. Relative heritability of skeletal nonmetric and metric traits. *Am. J. Phys. Anthropol.*, 59(2), 151–155.

Cho, Y.J. (2009). Radiological analysis of ponticulus posticus in Koreans. *Yonsei Med. J.*, 50(1), 45–49.

Chrismant, B. (2011). Le troisième trochanter chez les individus immatures. Site de Bruguières (Haute-Garonne, 31) VIIe–XIVe siècles. *Paleobios*, 16, 1–6.

Cooper, P.D., Stewart, J.H., McCormick, W.F. (1988). Development and morphology of the sternal foramen. *Am. J. Foren. Med. Path.*, 9(4), 342–347.

Coskun, N., Yuksel, M., Cevener, M., Arican, R.Y., Ozdemir, H., Bircan, O., Sindel, T., Ilgi, S., Sindel, M. (2009). Incidence of accessory ossicles and sesamoid bones in the feet: A radiographic study of the Turkish subjects. *Surg. Radiol. Anat.*, 31(1), 19–24.

Crubézy, E. (1988). Interactions entre facteurs bio-culturels, pathologie et caractères discrets. Exemple d'une population médiévale: Canac (Aveyron). Medical Thesis, Faculté de médecine, Montpellier.

Crubézy, E. (1991). Caractères discrets et évolution : exemple d'une population Nubienne : Missimia (Soudan). PhD Thesis, Université de Bordeaux, Bordeaux.

Crubézy, E. (1992). Caractères discrets et modifications de l'ontogenèse. *CR. Acad. Sci.*, 3, 235–239.

Crubézy, E. and Sellier, P. (1990). Caractères discrets et "recrutement" des ensembles sépulcraux. *Bull. Mem. Soc. Anthropol. Paris*, 2(3/4), 171–177.

Crubézy, E., Telmon, N., Sevin, A., Picard, J., Rougé, D., Larrouy, G., Braga, J., Ludes, B., Murail, P. (1999). Microévolution d'une population historique : étude des caractères discrets d'une population Missiminia (Soudan, IIIe–IVe siècles). *Bull. Mem. Soc. Anthropol. Paris*, 11 (1/2), 1–213.

Cunha, E. (2006). Pathology as a factor of personal identity in forensic anthropology. In *Forensic Anthropology and Medecine: Complementray Sciences from Recovery to Cause of Death*, Schmitt, A., Cunha, E., Pineiro, J. (eds). Humana Press, Totowa.

Curate, F. (2008). A case of *Os odontoideum* in the palaeopathological record. *Int. J. Osteoarchaeol.*, 18(1), 100–105.

Das, S.S., Mahajan, A., Vasudeva, N. (2016). Morphometric study of clavicular facet of coracoclavicular joint in adult Indian population. *J. Clin. Diagn. Res.*, 10(4), 8–11.

Dastugue, J. (1984). De l'appréciation des caractères dits "discrets" en anthropologie. *Bull. Mem. Soc. Anthropol. Paris*, 14(1), 237–244.

Debono, L. (2006). Évolution diachronique (anthropologique et paléopathologique) du squelette post-crânien des adultes d'une population historique du Sud-Est de la France (Nécropole Notre-Dame-du-Bourg, Digne, IVe–XVIIe siècle après J.-C.). PhD Thesis, Université Claude-Bernard, Lyon.

Desideri, J. and Eades, S. (2002). La question du "peuplement" campaniforme en Suisse occidentale : état de la question et apports de la morphologie dentaire et crânienne. *Bull. Mem. Soc. Anthropol. Paris*, 14(3/4), 345–362.

Dolensek, J., Cvetko, E., Snoj, Z., Meznaric, M. (2017). Complete occipitalization of the atlas with bilateral external auditory canal atresia. *Surg. Radiol. Anat.*, 39(9), 1053–1059.

Duncan, W. and Dahm, D.L. (2003). Clinical anatomy of the fabella. *Clin. Anat.*, 16(5), 448–449.

Duran, S.M., Gunaydin, E., Hatipoglu, H.G., Sakman, B. (2016). Bipartite patella: Magnetic resonance imaging. *Meandros Med. Dent. J.*, 17(1), 22–26.

Duray, S.M., Morter, H.B., Smith, F.J. (1999). Morphological variation in cervical spinous processes: Potential applications in the forensic identification of race from the skeleton. *J. For Sci.*, 44(5), 937–944.

Dutour, O. (1992). Activités physiques et squelette humain : le difficile passage de l'actuel au fossile. *Bull. Mem. Soc. Anthropol. Paris*, 4(3/4), 233–241.

Dutour, O. (2000). Chasse et activités physiques dans la Préhistoire : les marqueurs osseux d'activités chez l'homme fossile. *Anthropologie et Préhistoire*, 111, 156–165.

Egerci, O.F., Kose, O., Turan, A., Kilicaslan, O.F., Sekerci, R., Keles-Celik, N. (2017). Prevalence and distribution of the fabella: A radiographic study in Turkish subjects. *Folia Morphol.*, 76(3), 478–483.

El-Din, A.M.S. and El Banna, R.A.E. (2006). Congenital anomalies of the vertebral column: A case study on ancient and modern Egypt. *Int. J. Osteoarchaeol.*, 16(3), 200–207.

Fily, M., Crubézy, E., Ludes, B., Rougé, D., Midant-Reynes, B. (2001). Sternal perforation and bifids ribs. A possible familial case 5400 years old, an example of epigenetic control of development? *Bull. Mem. Soc. Anthropol. Paris*, 13(1/2), 5–13.

Finnegan, M. (1978). Nonmetric variation of infra-cranial skeleton. *J. Anat.*, 125, 23–37.

Gemmerich Pfister, I. (1999). Création d'une collection anthropologique de référence et application des caractères discrets dans le cas de généalogies connues. PhD Thesis, Université de Genève, Geneva.

Golthamer, C.R. (1957). Duplication of the clavicle (os subclaviculare). *Radiology*, 68(4), 576–578.

Gray, D. (1942). Variations in the human scapulae. *Am. J. Phys. Anthropol.*, 29(1), 57–72.

Grüneberg, H. (1952). Genetical studies on the skeleton of the mouse IV. Quasi-continuous variations. *J. Genet.*, 51(1), 95–114.

Haas, W.H., Kingma, M.J., Drucker, F. (1965). The coracoclavicular joint and related pathological conditions. *Ann. Rheum. Dis.*, 24, 257–266.

Hanihara, T. (1992). Dental and cranial affinities among populations of East Asia and the Pacific: The basic populations in East Asia, IV. *Am. J. Phys. Anthropol.*, 88(2), 163–182.

Hanihara, T., Ishida, H., Dodo, Y. (2003). Characterization of biological diversity through analysis of discrete cranial traits. *Am. J. Phys. Anthropol.*, 121(3), 241–251.

Hasan, M., Shukla, S., Siddiqui, M.S., Singh, D. (2001). Posterolateral tunnels and ponticuli in human atlas vertebrae. *J. Anat.*, 199, 339–343.

Hauser, G. and De Stefano, G.F. (1989). *Epigenetic Variants of the Human Skull*. Lubrecht & Cramer Ltd, Stuttgart.

Hosalkar, H.S., Gerardi, J.A., Shaw, B.A. (2001). Combined asymptomatic congenital anterior and posterior deficiency of the atlas. *Pediatr. Radiol.*, 31(11), 810–813.

Hrdlicka, A. (1934). The hypotrochanteric fossa of the femur. *Smithsonian Miscellaneous Collections*, 92(1), 1–49.

Huang, J., Zhang, Y., Ma, X., Wang, X., Zhang, C., Chen, L. (2014). Accessory navicular bone incidence in Chinese patients: A retrospective analysis of X-rays following trauma or progressive pain onset. *Surg. Radiol. Anat.*, 36(2), 167–172.

Hunt, D.R. and Bullen, L. (2007). The frequency of os acromiale in the Robert J. Terry collection. *Int. J. Osteoarchaeol.*, 17(3), 309–317.

Jit, I. and Kaur, H. (1986). Rhomboid fossa in the clavicles of north Indians. *Am. J. Phys. Anthropol.*, 70(1), 97–103.

Johnson, J. and Brogdon, B. (1982). Dorsal defect of the patella: Incidence and distribution. *Am. J. Roentgenol.*, 139(2), 339–340.

Kapetanakis, S., Giovannopoulou, E., Nastoulis, E., Demetriou, T. (2016). Butterfly vertebra. A case report and a short review of the literature. *Folia Morphol.*, 75(1), 117–121.

Karapetian, M.K. (2017). Discrete morphological variants of human cervical vertebrae: Exploring pattern of distribution and biological significance. *Homo*, 68(3), 176–198.

Kawashima, T., Takeishi, H., Yoshitomi, S., Ito, M., Sasaki, H. (2007). Anatomical study of the fabella, fabellar complex and its clinical implications. *Surg. Radiol. Anat.*, 29(8), 611–616.

Kennedy, K.A.R. (1989). Skeletal markers of occupational stress. In *Reconstruction of Life from the Skeleton*, Iscan, M.Y., Kennedy, K.A.R. (eds). Alan R. Liss Inc., New York.

Kim, M.S. (2015). Anatomical variant of atlas: Arcuate foramen, occpitalization of atlas, and defect of posterior arch of atlas. *J. Korean Neurosurg. S.*, 58(6), 528–533.

Kose, O., Eraslan, A., Ergun, A., Egerci, O.F., Ercan, E.C. (2015). Prevalence of bipartite patella in Turkish population. Analysis of bilateral knee radiographs in 897 subjects. *Int. J. Morphol.*, 33(3), 1108–1113.

Kostick, E. (1963). Facets and imprints on the upper and lower extremities of femora from a Western Nigerian population. *J. Anat.*, 97(3), 393–402.

Koyun, N., Aydinlioglu, A., Gumrukcuoglu, F.N. (2011). Aperture in coronoid-olecranon septum: A radiological evaluation. *Indian J. Orthop.*, 45(5), 392–395.

Kumar, G.R. and Mehta, C.D. (2008). A study of the incidence of supracondylar process of the humerus. *J. Anat. Soc. India*, 57(2), 111–115.

Kumar, J., Park, W.H., Kim, S.-H., Lee, H.I., Yoo, J.C. (2013). The prevalence of os acromiale in Korean patients visiting shoulder clinic. *Clin Orthop. Surg.*, 5(3), 202–208.

Lawson, J.P. (1994). Clinically significant radiologic anatomic variants of the skeleton. *Am. J. Roentgenol.*, 163(2), 249–255.

Le Double, A. (1903). *Traité des variations des os du crâne de l'Homme et de leur signification au point de vue de l'anthropologie zoologique.* Vigot Frères, Paris.

Le Double, A. (1906). *Traité des variations des os de la face de l'Homme et de leur signification au point de vue de l'anthropologie zoologique.* Vigot Frères, Paris.

Le Double, A. (1912). *Traité des variations de la colonne vertébrale de l'Homme et de leur signification au point de vue de l'anthropologie zoologique.* Vigot Frères, Paris.

Le Minor, J.M. (1987). Comparative anatomy and significance of the sesamoid bone of the peroneus longus muscle (os peroneum). *J. Anat.*, 151, 85–99.

Le Minor, J.M. (1997). The retrotransverse foramen of the human atlas vertebra – A distinctive variant within primates. *Acta Anat.*, 160(3), 208–212.

Lee, C.H., Park, C.M., Kim, K.A., Hong, S., Seol, H.Y., Kim, B.H., Kim, J.H. (2007). Identification and prediction of transitional vertebrae on imaging studies: Anatomical significance of paraspinal structures. *Clin. Anat.*, 20(8), 905–914.

Lim, C.K., Lee, K.W., Bin, J.C., Rhee, B.C. (1982). Congenital anomalies of the ribs. *J. Korean Radiol. Soc.*, 18(3), 487–495.

Macaluso, P.J. and Lucena, J. (2014). Morphological variations of the anterior thoracic skeleton and their forensic significance: Radiographic findings in a Spanish autopsy sample. *Forensic Sci. Int.*, 241, 220.e1-220.e7.

Mafart, B. (1980). *L'Abbaye Saint-Victor de Marseille, étude anthropologique de la nécropole des Ve et VIe siècles*. CNRS édition, Paris.

Mafart, B. (1984). Étude anthropologique de la nécropole paléochrétienne et médiévale de La Gayole (Var). PhD Thesis, Université de Provence, Marseille.

Mafart, B. (2005). Description, significance and frequency of the acetabular crease of the hip bone. *Int. J. Osteoarchaeol.*, 15(3), 208–215.

Mahato, N.K. (2010). Morphological traits in sacra associated with complete and partial lumbarization of first sacral segment. *Spine J.*, 10(10), 910–915.

Mann, R.W. (1990). Calcaneus secundarius: Description and frequency in six skeletal samples. *Am. J. Phys. Anthropol.*, 81(1), 17–25.

Masnicova, S. and Benus, R. (2003). Developmental anomalies in skeletal remains from the Great Moravia and Middle Ages cemeteries at Devin (Slovakia). *Int. J. Osteoarchaeol.*, 13(5), 266–274.

Mellado, J.M., Ramos, A., Salvado, E., Camins, A., Danus, M., Sauri, A. (2003). Accessory ossicles and sesamoid bones of the ankle and foot: Imaging findings, clinical significance and differential diagnosis. *Eur. Radiol.*, 13, 164–177.

Merbs, C.F. (2004). Sagittal clefting of the body and other vertebral development errors in Canadian Inuit skeletons. *Am. J. Phys. Anthropol.*, 123(3), 236–249.

Miki, T., Oka, M., Urushidani, H., Hirofuji, E., Tanaka, S., Iwamoto, S. (1979). Ponticulus posticus: Its clinical significance. *Acta Med. Kinki Univ.*, 4(2), 427–430.

Miles, A.E.W. (1996). Humeral impingement on the acromion in a Scottish Island population of c.1600 AD. *Int. J. Osteoarchaeol.*, 6(3), 259–288.

Mitchell, J. (1998). The incidence of the lateral bridge of the atlas vertebra. *J. Anat.*, 193, 283–285.

Miyakoshi, N., Kobayashi, A., Hongo, M., Shimada, Y. (2015). Sacral rib: An uncommon congenital anomaly. *Spine J.*, 15(6), 35–38.

Murail, P. (1996). Biologie et pratiques funéraires des populations d'époque historique : une démarche méthodologique appliquée à la nécropole gallo-romaine de Chantambre (Essone, France). PhD Thesis, Université Bordeaux 1, Bordeaux.

Murail, P. (2005). Variations anatomiques non métriques : les caractères discrets. In *Objets et méthodes en paléoanthropologie*, Dutour, O., Hublin, J.-J., Vandermeersch, B. (eds). CHTS Édition, Paris.

Nakajima, A., Usui, A., Hosokai, Y., Kawasumi, Y., Abiko, K., Funayama, M., Saito, H. (2014). The prevalence of morphological changes in the thoracolumbar spine on whole-spine computed tomographic images. *Insights into Imaging*, 5(1), 77–83.

Natsis, K. (2008). Supracondylar process of the humerus: Study on 375 Caucasian subjects in Cologne, Germany. *Clin. Anat.*, 21(2), 138–141.

Natsis, K., Totlis, T., Tsikaras, P., Appell, H.J., Skandalakis, P., Koebke, J. (2007). Proposal for classification of the suprascapular notch: A study on 423 dried scapulas. *Clin. Anat.*, 20(2), 135–139.

Natsis, K., Totlis, T., Chorti, A., Karanassos, M., Didagelos, M., Lazaridis, N. (2016). Tunnels and grooves for supraclavicular nerves within the clavicle: Review of the literature and clinical impact. *Surg. Radiol. Anat.*, 38(6), 687–691.

Nayak, S.R., Das, S., Krishnamurthy, A., Prabhu, L.V., Potu, B.K. (2009). Supratrochlear foramen of the humerus: An anatomico-radiological study with clinical implications. *Upsala J. Med. Sci.*, 114(2), 90–94.

O'Brien, J., Murphy, C., Halpenny, D., McNeill, G., Torreggiani, W.C. (2011). Magnetic resonance imaging features of asymptomatic bipartite patella. *Eur. J. Radiol.*, 78(3), 425–429.

Odgers, P.N.B. (1931). Two details about the neck of the femur: (1) The eminentia. (2) The empreinte. *J. Anat.*, 65, 352–362.

Offenbecker, A.M. and Case, D.T. (2012). Accessory navicular: A heritable accessory bone of the human foot. *Int. J. Osteoarchaeol.*, 22(2), 158–167.

Ossenberg, N. (1969). Discontinuous morphological variation in the human cranium. PhD Thesis, University of Toronto, Toronto.

Ossenberg, N. (1994). Origins and affinities of the native people of the northwestern North America: The evidence of cranial nonmetric traits. In *Method and Theory for Investigating the Peopling of the Americas*, Bonnichsen, R., Steele, D.G. (eds). The Center for the Study of the First American, Oregon State University, Oregon.

Oyedele, O., Maseko, C., Mkasi, N., Mashanyana, M. (2006). High incidence of the os peroneum in a cadaver sample in Johannesburg, South Africa: Possible clinical implications? *Clin. Anat.*, 19(7), 605–610.

Ozdolap, S., Sarikaya, S., Balbaloglu, O., Kalayci, M. (2007). Congenital defects of posterior arch of the atlas: A case report. *Neuroanatomy*, 6, 72–74.

Pais, J., Levine, A., Pais, S.O. (1978). Coccygeal ribs: Development and appearance in two cases. *Am. J. Roentgenol.*, 131(1), 164–166.

Paraskevas, G., Papaziogas, B., Tsonidis, C., Kapetanos, G. (2005). Gross morphology of the bridges over the vertebral artery groove on the atlas. *Surg. Radiol. Anat.*, 27(2), 129–136.

Perdikakis, E., Grigoraki, E., Karantanas, A. (2011). Os naviculare: The multi-ossicle configuration of a normal variant. *Skeletal Radiol.*, 40(1), 85–88.

Perez, I.E., Chavez, A.K., Ponce, D. (2014). Frequency of ponticulus posticus in lateral cephalometric radiography of Peruvian patients. *Int. J. Morphol.*, 32(1), 54–60.

Polguj, M., Jedrzejewski, K., Podgorski, M., Topol, M. (2011). Morphometric study of the suprascapular notch: Proposal of classification. *Surg. Radiol. Anat.*, 33(9), 781–787.

Polguj, M., Podgorski, M., Jedrzejewski, K., Topol, M. (2012). The double suprascapular foramen: Unique anatomical variation and the new hypothesis of its formation. *Skeletal Radiol.*, 41(12), 1631–1636.

Radi, N., Mariotti, V., Riga, A., Zampetti, S., Villa, C., Belcastro, M.G. (2013). Variation of the anterior aspect of the femoral head-neck junction in a modern human identified skeletal collection. *Am. J. Phys. Anthropol.*, 152(2), 261–272.

Rashid, M., Khalid, M., Malik, N. (2008). Sacral rib: A rare congenital anomaly. *Acta Orthop. Belg.*, 74(3), 429–431.

Riddick, L., Brogdon, B.G., Lasswellhoff, J., Delmas, B. (1983). Radiographic identification of charred human remains through use of the dorsal defect of the patella. *J. Forensic Sci.*, 28(1), 263–267.

Robson, K.A. (2011). Os odontoideum: Rare cervical lesion. *West. J. Emerg. Med.*, 12(4), 520–522.

Saikia, R., Baishya, R.J., Deka, B. (2017). Variations in the shape of the suprascapular notch in dry human scapula: An anatomical study. *Int. J. Sci. Study*, 5(1), 187–190.

Saini, V., Singh, R., Bandopadhyay, M., Tripathi, S., Sharma, S. (2009). Occipitalization of the atlas: Its occurrence and embryological basis. *International Journal of Anatomical Variations*, 2, 65–68.

Saluja, P.G. (1988). The incidence of spina-bifida occulta in a historic and a modern London population. *J. Anat.*, 158, 91–93.

Sammarco, V.J. (2000). Os acromiale: Frequency, anatomy, and clinical implications. *J. Bone Joint Surg. Am.*, 82(3), 394–400.

Saunders, S. (1978). The development and distribution of discontinuous morphological variation of the human infracranial skeleton. PhD Thesis, University of Toronto, Toronto.

Saunders, S. (1989). Non-metric skeletal variation. In *Reconstruction of Life from the Skeleton*, Iscan, M.Y., Kennedy, K.A.R. (eds). Wiley-Liss, New York.

Saunders, S. and Popovich, F. (1978). A family study of two skeletal variants: Atlas bridging and clinoid bridging. *Am. J. Phys. Anthropol.*, 49, 193–204.

Saupe, N., Mengiardi, B., Pfirrmann, C.W.A., Vienne, P., Seifert, B., Zanetti, M. (2007). Anatomic variants associated with peroneal tendon disorders: MR imaging findings in volunteers with asymptomatic ankles. *Radiol.*, 242(2), 509–517.

Scheuer, L. and Black, S. (2004). *The Juvenile Skeleton*. Academic Press, London.

Scheuer, L., Black, S., Schaefer, M. (2009). *Juvenile Osteology: A Laboratory and Field Manual*. Academic Press, London.

Schilling, J., Schilling, A., Galdames, I.S. (2010). Ponticulus posticus on the posterior arch of atlas, prevalence analysis in asymptomatic patients. *Int. J. Morphol.*, 28(1), 317–322.

Scott, G. and Turner, C. (1997). *The Anthropology of Modern Human Teeth: Dental Morphology and Its Variation in Recent Human Populations*. Cambridge University Press, Cambridge.

Senoglu, M., Safavi-Abbasi, S., Theodore, N., Bambakidis, N.C., Crawford, N.R., Sonntag, V.K.H. (2007). The frequency and clinical significance of congenital defects of the posterior and anterior arch of the atlas. *J. Neurosurg-Spine*, 7(4), 399–402.

Singh, I. (1959). Squatting facets on the talus and tibia in Indians. *J. Anat.*, 93, 540–550.

Singh, J. and Pathak, R.K. (2013). Sex and age related non-metric variation of the human sternum in a Northwest Indian postmortem sample: A pilot study. *Forensic Sci. Int.*, 228(1/3), 181.e1–181.e12.

Sjovold, T. (1973). The occurrence of minor, non-metrical variants in the skeleton and their quantitative treatment for population comparisons. *Homo*, 24, 204–233.

Sjovold, T. (1984). A report on the heritability of some cranial measurements and non-metric traits. In *Multivariale Statistical Methods in Physical Anthropology*, van Vark, G.N., Howells, W.W. (eds). Springer, Dordrecht.

Slouma, M., Cheour, E., Sahli, H., Elleuch, M. (2016). Congenital butterfly vertebra. *Acta Med. Port.*, 29(7/8), 491.

Sonel, B., Yalcin, P., Ozturk, E.A., Bokesoy, I. (2001). Butterfly vertebra: A case report. *Clin. Imag.*, 25(3), 206–208.

Song, W.C., Kim, S.H., Park, D.K., Koh, K.S. (2009). Bifid Rib: Anatomical considerations in three cases. *Yonsei Med. J.*, 50(2), 300–303.

Stewart, T. (1953). The age incidence of neural-arch defects in Alaskan natives, considered from the standpoint of etiology. *J. Bone Joint Surg.*, 35(4), 937–950.

Tague, R.G. (2011). Sacralization is not associated with elongated cervical costal process and cervical rib. *Clin. Anat.*, 24(2), 209–217.

Tague, R.G. (2018). Proximate cause, anatomical correlates, and obstetrical implication of a supernumerary lumbar vertebra in humans. *Am. J. Phys. Anthropol.*, 165(3), 444–456.

Takeuchi, M., Aoyama, M., Wakao, N., Tawada, Y., Kamiya, M., Osuka, K., Matsuo, N., Takayasu, M. (2016). Prevalence of C7 level anomalies at the C7 level: An important landmark for cervical nerve ultrasonography. *Acta Radiol.*, 57(3), 318–324.

Thoma, A. (1981). The pattern of quasi-continuous variation in Homo sapiens. *J. Hum. Evol.*, 10, 303–310.

Thomas, A. (2011). Identités funéraires, variants biologiques et facteurs chronologiques : une nouvelle perception du contexte culturel et social du Cerny (Néolithique moyen I du Bassin parisien). PhD Thesis, Université Bordeaux 1, Bordeaux.

Trotter, M. (1937). Accessory sacroiliac articulations. *Am. J. Phys. Anthropol.*, 12(2), 247–261.

Turkay, R., Inci, E., Ors, S., Nalbant, M.O., Gurses, I.A. (2017). Frequency of sternal variations in living individuals. *Surg. Radiol. Anat.*, 39(11), 1273–1278.

Twigg, H.L. and Rosenbaum, R.C. (1981). Duplication of the clavicle. *Skeletal Radiol.*, 6(4), 281–281.

Uygur, E., Aktas, B., Oz, T.T., Erinc, S., Koc, M. (2016). Prevalence of accessory bones of the foot in Turkish patients. *J. Acad. Res. Med.*, 6(3), 147–151.

Verna, E. (2014). Les variations osseuses asymptomatiques du squelette postcrânien : leur contribution à l'identification en anthropologie médico-légale. PhD Thesis, Université d'Aix-Marseille, Marseille.

Verna, E. (2016). Les variations osseuses asymptomatiques du squelette postcrânien humain : leur contribution à l'identification en anthropologie médicolégale. *Bull. Mem. Soc. Anthropol. Paris*, 28, 226–232.

Verna, E., Piercecchi-Marti, M-D., Chaumoitre, K., Panuel, M., Adalian, P. (2014). Mise au point sur les caractères discrets du membre inférieur : définition, épidémiologie, étiologies. *Bull. Mem. Soc. Anthropol. Paris*, 26, 52–66.

Verna, E., Piercecchi-Marti, M-D., Chaumoitre, K., Adalian, P. (2015). Relevance of discrete traits in forensic anthropology: From the first cervical vertebra to the pelvic girdle. *Forensic Sci. Int.*, 253, 134.e1–134.e7.

Viertel, V.G., Intrapiromkul, J., Maluf, F., Patel, N.V., Zheng, W., Alluwaimi, F., Walden, M.J., Belzberg, A., Yousem, D.M. (2012). Cervical ribs: A common variant overlooked in CT imaging. *Am. J. Neuroradiol.*, 33(11), 2191–2194.

Villotte, S. (2008). Les marqueurs ostéoarticulaires d'activité. In *Ostéo-archéologie et techniques médico-légales : tendances et perspectives*, Charlier, P. (ed.). De Boccard, Paris.

Willis, T.A. (1929). An analysis of vertebral anomalies. *Am. J. Surg.*, 6(2), 163–168.

Wilson, T.C. (2006). Duplication of the clavicle. *Orthopedics*, 29(1), 79–80.

Wood-Jones, F. (1931). The non-metric morphological characters of the skull as criteria for racial diagnosis. I, II, III. *J. Anat.*, 65, 179–195, 368–378, 438–445.

Wood-Jones, F. (1934). The non-metric morphological characters of the skull as criteria for racial diagnosis. IV, V. *J. Anat.*, 68, 96–108, 323–330.

Wu, L.P., Li, Y.K., Li, Y.M., Zhang, Y.Q., Zhong, S.Z. (2009). Variable morphology of the sacrum in a Chinese population. *Clin. Anat.*, 22(5), 619–626.

Wysocki, J., Bubrowski, M., Reymond, J., Kwiatkowski, J. (2003). Anatomical variants of the cervical vertebrae and the first thoracic vertebra in man. *Folia Morphol.*, 62(4), 357–363.

Yekeler, E., Tunaci, M., Tunaci, A., Dursun, M., Acunas, G. (2006). Frequency of sternal variations and anomalies evaluated by MDCT. *Am. J. Roentgenol.*, 186(4), 956–960.

Index

L, N

O, P

R, S, T

Other titles from

iSTE

in

Biology and Biomedical Engineering

2022

DAMBRICOURT MALASSÉ Anne
Embryogeny and Phylogeny of the Human Posture 2: A New Glance at the Future of our Species (Comparative Anatomy and Posture of Animal and Human Set – Volume 4)

FOSSÉ Philippe
Structures and Functions of Retroviral RNAs: The Multiple Facets of the Retroviral Genome (Nucleic Acids Set - Volume 1)

2021

DAMBRICOURT MALASSÉ Anne
Embryogeny and Phylogeny of the Human Posture 1: A New Glance at the Future of our Species (Comparative Anatomy and Posture of Animal and Human Set – Volume 3)

GRANDCOLAS Philippe, MAUREL Marie-Christine
Systematics and the Exploration of Life

HADJOUIS Djillali
The Skull of Quadruped and Bipedal Vertebrates: Variations, Abnormalities and Joint Pathologies (Comparative Anatomy and Posture of Animal and Human Set – Volume 2)

HULLÉ Maurice, VERNON Philippe
The Terrestrial Macroinvertebrates of the Sub-Antarctic Îles Kerguelen and Île de la Possession

2020

CAZEAU Cyrille
Foot Surgery Viewed Through the Prism of Comparative Anatomy: From Normal to Useful
(Comparative Anatomy and Posture of Animal and Human Set – Volume 1)

DUJON Bernard, PELLETIER Georges
Trajectories of Genetics

2019

BRAND Gérard
Discovering Odors

BUIS Roger
Biology and Mathematics: History and Challenges

2016

CLERC Maureen, BOUGRAIN Laurent, LOTTE Fabien
Brain-Computer Interfaces 1: Foundations and Methods
Brain-Computer Interfaces 2: Technology and Applications

FURGER Christophe
Live Cell Assays: From Research to Health and Regulatory Applications

2015

CLARYSSE Patrick, FRIBOULET Denis
Multi-modality Cardiac Imaging: Processing and Analysis

2014

CHEZE Laurence
Kinematic Analysis of Human Movement

CLARYSSE Patrick, FRIBOULET Denis
Multi-modality Cardiac Imaging: Processing and Analysis

DAO Tien Tuan, HO BA THO Marie-Christine
Biomechanics of the Musculoskeletal: Modeling of Data Uncertainty and Knowledge

FANET Hervé
Medical Imaging Based on Magnetic Fields and Ultrasounds

FARINAS DEL CERRO Luis, INOUE Katsumi
Logical Modeling of Biological Systems

MIGONNEY Véronique
Biomaterials

TEBBANI Sihem, LOPES Filipa, FILALI Rayen, DUMUR Didier, PAREAU Dominique
CO_2 Biofixation by Microalgae: Modeling, Estimation and Control

Printed and bound by CPI Group (UK) Ltd, Croydon, CR0 4YY
16/09/2026
14963081-0001